Bi & Gi Publishers

Current Topics in Rehabilitation
Series Editor : R. Corsico M.D.

Titles in the series:

Respiratory Muscles in Chronic Obstructive Pulmonary Disease
Edited by: A. Grassino, C. Fracchia, C. Rampulla, L. Zocchi

Pathophysiology and Treatment of Pulmonary Circulation
Edited by: A. Morpurgo, R. Tramarin, C. Rampulla, C. Fracchia, F. Cobelli

Chronic Pulmonary Hyperinflation
Edited by: A. Grassino, C. Rampulla, N. Ambrosino, C. Fracchia

Biochemistry of Pulmonary Emphysema
Edited by: C. Grassi, J. Travis, L. Casali, M. Luisetti

Right Ventricular Hypertrophy and Function in Chronic Lung Disease
Edited by: V. Jĕzeck, M. Morpurgo, R. Tramarin

Forthcoming titles in the series:

Pulmonary Rehabilitation
Edited by: A. Grassino, C. Rampulla, C. Fracchia

Acknowledgments: The Organization Committee of the Workshop on "Nutrition and Ventilatory Function" held in Salice Terme (Pavia) 9-11 November 1989 gratefully acknowledges the Camillo Corvi S.p.A. Piacenza (Italy) for the support and co-operation.

Nutrition and Ventilatory Function

Edited by: R. D. Ferranti, C. Rampulla
C. Fracchia, N. Ambrosino

Foreword by: R. Corsico
With 49 figures and 39 tables

Springer-Verlag London Ltd.

R. D. Ferranti, Center for Breathing Disorders, Gaylord Hospital, Yale University School of Medicine, USA

C. Rampulla, Clinica del Lavoro Foundation, Institute of Care and Research, Medical Center of Rehabilitation, Montescano (Pavia) Italy

C. Fracchia, Clinica del Lavoro Foundation, Institute of Care and Research, Medical Center of Rehabilitation, Montescano (Pavia) Italy

N. Ambrosino,Clinica del Lavoro Foundation, Institute of Care and Research, Medical Center of Rehabilitation, Montescano (Pavia) Italy

Series Editor:

R. Corsico - Clinica del Lavoro Foundation, Institute of Care and Research, Medical Center of Rehabilitation, Montescano (Pavia) Italy

ISBN 978-1-4471-3842-6   ISBN 978-1-4471-3840-2 (eBook)
DOI 10.1007/978-1-4471-3840-2

British Library Cataloguing in Publication Data
Nutrition and Ventilatory Function (Current Topics in Rehabilitation Series)
I. Ferranti, Redento D. II. Series 616.2

Library of Congress Cataloging in Publication Data
Nutrition and ventilatory function / edited by R.D. Ferranti... (et al.); foreword by R. Corsico
p. cm. — (Current topics in rehabilitation) Includes index

1. Lungs Diseases, Obstructive—Nutritional aspects.
2. Malnutrition. I. Ferranti, R.D. (Redento D.). 1923-II. Series.
(DNLM: 1. Lung—physiopathology. 2. Lung Diseases, Obstructive.
3. Nutrition. WF 600 N976) RC776.03N87 1992 616.2 '4—dc20
DNLM/DLC for Library of Congress 92-2313 CIP

© 1992 **Springer-Verlag London**
Originally published by Bi & Gi publishers in 1992
Softcover reprint of the hardcover 1st edition 1992

Foreword

In his review of the Proceedings of the First International Workshop on *Respiratory Muscles in Chronic Obstructive Pulmonary Disease*, organized in Montescano in 1986, Thomas K. Aldrich stated that "there is little information on nutrition".

We felt that the need to supplement the evaluation of respiratory muscles by studying nutritional problems was an important enough matter to justify producing a specific study on the subject. This book is the result of the efforts made in this direction.

Since 1986, many of the problems associated with nutrition and ventilatory function have been elucidated, and we hope that the contents of this book may contribute towards spreading interest and knowledge in this sector, particularly among those working in the field of respiratory rehabilitation.

The physiological and clinical aspects of nutritional problems and the therapeutic approaches adopted in different disease conditions are dealt with in these pages by some of the best-known researchers in the field, providing abundant evidence, if any should be needed, that rehabilitation is not merely exercise. The picture strongly emerging over these last few years is that rehabilitation is not merely a matter of knowledge of lung or respiratory muscles, but also of heart-lung interaction, metabolic equilibrium and nutrition. So much for recent developments, but what does the future hold in store?

May 1992

RENATO CORSICO

Preface

Nutrition and ventilation are inextricably related. Oxygen and nutrients are necessary and participate together in the process of respiration and in furnishing the energy necessary for the activities of daily living.

The focus of this book is on the relationship between these two functions. The purpose is an attempt to reach a better understanding of how ventilatory disorders may affect the nutritional status of the patient, or can be affected by nutritional strategies or other interventions which impact on energy balance.

Special emphasis is on dysfunction from the increased work of breathing as it is manifested in obesity and COPD, in other conditions such as cystic fibrosis, and on hypercatabolic states as in the respiratory distress syndrome and in multiorgan failure. The effects of malnutrition on the immunological status of the patient are also highlighted.

While much of the material presented can be easily translated into clinical protocols, the contributions present stimulating ideas and questions which need further elaboration and answers.

In COPD, there is a significant correlation between the increase in the work of breathing due to the obstructive condition and the nutritional status of the patient, but other cofactors appear to also play a significant role.

More investigation at the molecular level is needed to better assess the role of certain products of the inflammatory response in promoting cachexia, and on the possibility that pharmacological agents could alter it. Micronutrients may also offer more application in therapy, as they are necessary cofactors in antioxidant enzyme systems and as free radical scavengers.

In assessing the nutritional effect on ventilatory function, composition and quantity of diet will be considered. Systems providing content and transport of oxygen and nutrients and regulations of CO_2, electrolytes and metabolic balance are complex and may fail at any point. Requirements vary among diseases, disease stages and individuals. In hypermetabolic conditions such as COPD, an increase in caloric intake is needed. In other conditions such as neuromuscular insufficiency, overfeeding my result in excess fat deposits and increased respiratory work.

In specific conditions, nutritional strategies can benefit the therapeutic regimen. Reducing carbohydrates in favor of proteins and fats in acid maltase deficiency, which also affects the ventilatory muscles, may reduce dependence on a ventilator. In the treatment of COPD and in weaning patients from ventilatory assistance, lowering the carbohydrate to fat ratio minimizes the increase in Respiratory Quotient and work of breathing.

When refeeding is needed, the enteral route is preferred. This appears to avoid mucosal atrophy and prevent bacterial translocation across the intestinal mucosa as a source of systemic sepsis and multiorgan failure, as well as reducing production of inflammatory mediators from the liver. Glutamine, short chain fatty acids and ketones, in the diet, appear to protect vascular endothelium and parenchymal damage. In the lung, this damage produces respiratory failure.

In hypercatabolic states, there is a tendency to retain sodium and fluids. "Overload" of carbohydrates and fat may not be well tolerated. Nutrition here is essential to maintain a metabolic equilibrium and prevent loss of visceral proteins, but a positive metabolic balance aimed at the restoration of muscle mass and independence from mechanical ventilation may need to be deferred until the acute process subsides.

Specific conditioning exercises for strength and endurance improve performance in ventilatory muscles. The interaction between muscle conditioning, the nutritional and metabolic status, and the oxygenation of the patient, remain of critical importance and interest.

This book provides an update review of these subjects and indicates the many areas in need of further investigation. We are grateful to the contributors for presenting these most provocative works and thoughts.

R.D. Ferranti

Contents

Contributors

AMBROSINO N.
Division of Pulmonary Diseases, Clinica del Lavoro Foundation, Institute of Care and Research, Medical Center of Rehabilitation, Montescano (Pavia) Italy

ANTENORI S.
Department of Respiratory Physiopathology, "A. Murri" Hospital, Jesi, Ancona, Italy

ASKANAZI J.
Division of Critical Care Medicine, Department of Anesthesiology, Albert Einstein College of Medicine, Montefiore Medical Center, New York, USA

BONANDRINI L.
Chair of Microsurgery, University of Pavia, Italy

BORGHETTI A.
Internal Medicine and Nephrology, University of Parma, Italy

BRAUN N. M. T.
Department of Medicine, Pulmonary Division, St. Luke's-Roosevelt Hospital Center, Clinical Medicine, College of Physicians and Surgeons, Columbia University, New York, USA

BRAUN S.R.
Pulmonary, Critical Care and Environmental Medicine, University of Missouri, Columbia, USA

BROWN C. III
Pulmonary Section, Gaylord Hospital, State of Connecticut, USA

CALLEGARI S.
Psychology Unit, Clinica del Lavoro Foundation, Institute of Care and Research, Medical Center of Rehabilitation, Montescano (Pavia) Italy

CALVI A.
Nutritional Unit, Clinica del Lavoro Foundation, Institute of Care and Research, Medical Center of Rehabilitation, Montescano (Pavia) Italy

CAPELLI A.
Division of Pulmonary Diseases, Clinica del Lavoro Foundation, Institute of Care and Research, Medical Center of Rehabilitation, Veruno (Novara), Italy

CEBRELLI T.
Department of Surgery, Istituto di Patologia Chirurgica, University of Pavia, Italy

COELHO C. A.
Communication Disorders Section, Gaylord Hospital, State of Connecticut, USA

COFFRINI E.
Internal Medicine and Nephrology, University of Parma, Italy

D'ALOYA N.
Department of Respiratory Physiopathology, "A. Murri" Hospital, Jesi, Ancona, Italy

DIONIGI P.
Department of Surgery, Istituto di Patologia Chirurgica, University of Pavia, Italy

DONNER C. F.
Division of Pulmonary Diseases, Clinica del Lavoro Foundation, Institute of Care and Research, Medical Center of Rehabilitation, Veruno (Novara), Italy

FERRANTI R. D.
Center for Breathing Disorders, Gaylord Hospital, Yale University School of Medicine, USA

FIACCADORI E.
Internal Medicine and Nephrology, University of Parma, Italy

FITTING J. W.
Division of Pneumology, Internal Medicine Department, Hospital University Centre Vaudois, Lausanne, Switzerland

FRACCHIA C.
Division of Pulmonary Diseases, Clinica del Lavoro Foundation, Institute of Care and Research, Medical Center of Rehabilitation, Montescano (Pavia) Italy

FRASCAROLO P.
Institute of Physiology, Faculty of Medicine, Lausanne University, Switzerland

GOLDSTEIN-SHAPSES S. A.
Department of Orthopaedic Surgery, Columbia Presbyterian Medical Center, New York, USA

GONZI G.
Internal Medicine and Nephrology, University of Parma, Italy

GRASSINO A.
Meakins-Christie Laboratory, Royal Victoria Hospital, McGill University and Notre Dame Hospital, University of Montreal, Quebec, Canada

GROSS D.
Department of Anesthesiology, Hadassah University Hospital, Jerusalem, Israel

IAPICHINO G.
ICU "E. Vecla" Institute of Anesthesiology and Intensive Care, University of Milan, IRCCS, Ospedale Maggiore, Milan, Italy

JEMOS V
Department of Surgery, Istituto di Patologia Chirurgica, University of Pavia, Italy

JÉQUIER E.
Institute of Physiology, Faculty of Medicine, Lausanne University, Switzerland

KATZ D. P.
Division of Critical Care Medicine, Department of Anesthesiology, Albert Einstein College of Medicine, Montefiore Medical Center, New York, USA

KINNEY J. M
Visiting Professor and Physician, Rockefeller University, Attending in Medicine and Surgery, St. Luke's-Roosevelt Hospital Center, New York, USA

KIRVELÄ O.
Division of Critical Care Medicine, Department of Anesthesiology, Albert Einstein College of Medicine, Montefiore Medical Center, New York, USA

LENTINI L.
Department of Respiratory Physiopathology, "A. Murri" Hospital, Jesi, Ancona, Italy

LUSUARDI M.
Division of Pulmonary Diseases, Clinica del Lavoro Foundation, Institute of Care and Research, Medical Center of Rehabilitation, Veruno (Novara), Italy

MAIANI G.
Psychology Unit, Clinica del Lavoro Foundation, Institute of Care and Research, Medical Center of Rehabilitation, Montescano (Pavia) Italy

MEYRON E.
Department of Dietary Services, Hadassah University Hospital, Jerusalem, Israel

MONTAGNA T.
Division of Pulmonary Diseases, Clinica del Lavoro Foundation, Institute of Care and Research, Medical Center of Rehabilitation, Montescano (Pavia) Italy

MURRAY M. J.
Critical Care Service, Department of Anesthesiology, Nutritional Support Service, Department of Internal Medicine, Mayo Clinic and Mayo Foundation Rochester, Minnesota, USA

NIEDERMAN M. S.
Medical and Respiratory Intensive Care Unit, Winthrop University Hospital, SUNY, Stony Brook, New York, USA

NØRREGAARD O.
Department of Respiratory Medicine, Aarhus University Hospital, Denmark

RAMPULLA C.
Respiratory Physiopathology Unit, Clinica del Lavoro Foundation, Institute of Care and Research, Medical Center of Rehabilitation, Montescano (Pavia) Italy

ROBERTO M.
Department of Health Services, State of Connecticut, USA

RONDA N.
Internal Medicine and Nephrology, University of Parma, Italy

SAMPSON M. G.
Department of Medicine, Division of Pulmonary Disease, State University of New York, Stony Brook, New York, USA
SCHINDLER D.
Maccabi Sick Fund, Tel-Aviv, Israel
SCHOLS A. M. W. J.
Department of Pulmonary Diseases, University of Limburg, Maastricht, Netherlands
SCHUTZ Y.
Institute of Physiology, Faculty of Medicine, Lausanne University, Switzerland
SKEIE B.
Division of Critical Care Medicine, Department of Anesthesiology, Albert Einstein College of Medicine, Montefiore Medical Center, New York, USA
SØREIDE E.
Division of Critical Care Medicine, Department of Anesthesiology, Albert Einstein College of Medicine, Montefiore Medical Center, New York, USA
SPADA E. L.
Division of Pulmonary Diseases, Clinica del Lavoro Foundation, Institute of Care and Research, Medical Center of Rehabilitation, Veruno (Novara), Italy
SUBIACO S.
Department of Respiratory Physiopathology, "A. Murri" Hospital, Jesi, Ancona, Italy
VIOLA L.
Psychology Unit, Clinica del Lavoro Foundation, Institute of Care and Research, Medical Center of Rehabilitation, Montescano (Pavia) Italy
WOUTERS E. F. W.
Department of Pulmonary Diseases, University of Limburg, Maastricht, Netherlands
ZACCARIA S.
Division of Pulmonary Diseases, Clinica del Lavoro Foundation, Institute of Care and Research, Medical Center of Rehabilitation, Veruno (Novara), Italy
ZANABONI S.
Division of Pulmonary Diseases, Clinica del Lavoro Foundation, Institute of Care and Research, Medical Center of Rehabilitation, Veruno (Novara), Italy

Nutrition and Malnutrition

1. Anthropometric Parameters in the Assessment of Nutritional Status

P. FRASCAROLO, Y. SCHUTZ, E. JÉQUIER

Institute of Physiology, Faculty of Medicine, Lausanne University, Switzerland

The nutritional status of a patient can be evaluated with a combination of laboratory tests and body composition assessment. In contrast to the assessment of vitamins and mineral status, the assessment of body composition cannot be inferred from simple biochemical tests. Numerous methods have been developed to assess body composition in man. Due to their high complexity, certain methods cannot be applied under clinical conditions (Table I). This includes for example densitometry. This paper will focus on anthropometric measurements (i.e. comparative body measurements) and bioelectrical impedance primarily used in clinical conditions to evaluate the various compartments of the organism.

Since excess body weight is in most instances accompanied with an excess

Table I. Principal methods used to assess body composition in man.

Methods	Application
Total body water	Research
Total body potassium	Research
Densitometry	Research
Anthropometry	Clinical + Research
Neutron activation analysis	Research
Muscle metabolites	Clinical + Research
Absorptionometry	Research
Bioelectrical conductance	Clinical + Research
Computerized tomography	Clinical + Research
Magnetic resonance imaging	Clinical + Research

adipose tissue storage, body weight measurements in relation to height, i.e. weight for height, can be indirectly used to infer about excessive amounts of adipose tissue. However this is by no means an evaluation of body fat "per se" since weight for height is partly dependent upon the body frame of the individual and this does not take into account the body composition (i.e. the fat to fat-free mass ratio) of the excess weight. Similarly, a drastically reduced body weight for height is an indirect indication of decreased amount of adipose tissue but this gives no indication of body composition. In epidemiological studies the measurement of body weight has provided an enormous amount of information about relative degrees of overweight in various populations. The most widely used reference table has been published by the Metropolitan Life Insurance Company.[6]

The desirable weight for a given height is given for both sexes and for different frame sizes. "Desirable" weight or so-called "ideal" body weight (IBW) corresponds to a weight for a given height at which the mortality rate has been found to be minimal. Unfortunately, no objective criteria have been given for the determination of frame size. Generally, the average weight for the middle frame size is used as a reference point. The main advantage of utilizing weight tables to assess the degree of overweight and underweight is its simplicity. However it should be recalled that in the Table of the Metropolitan Life Insurance, in many instances weight and height were not directly measured but were obtained anamnestically. When assessed the weight was measured with clothes and height was measured with shoes so that further corrections were found to be necessary. In addition, the revised Table of the Metropolitan Life Insurance,[7] gives divergent results, as compared to the early Table,[6] especially for short individuals.

At the end of the 19th century, the Belgian astronomist Quetelet (1869)[8] suggested an empirical index to calculate excess body weight, the so-called "Quetelet" index, later called body mass index (BMI). It is simply the ratio between the body weight (expressed in Kg) divided by height squared (expressed in m^2). It should be stressed that the height squared (m^2) does not represent the body surface area. As shown in figure 1, the normal range for the body mass index is 20 to 25 Kg/m^2 irrespective of the sex.[2] The degree of overweight or obesity has been classified into 3 categories: grade I (overweight or moderately obese), grade II (obese), and grade III (grossly obese). Up to recent times, the body mass index was used primarily to assess the degree of obesity in population studies. This obviously has limited value for the clinician who is interested in a given individual and who may be confronted with underweight rather than overweight patients. Recently, James et al.[3] have attempted to use the body mass index (BMI) for assessing the magnitude of underweight individuals subjected to chronic energy deficiency. They developed different cut-off points for BMI below 20. Three cut-off points for BMI were suggested: 18.5, 17 and 16 kg/m^2 (Fig. 1). A BMI between 18.5 and 20 was classified as "normal" although the person can be judged as lean. Diagnosis of grade I vs

$$BMI = \frac{WEIGHT}{HEIGHT^2} \quad (Kg/m2)$$

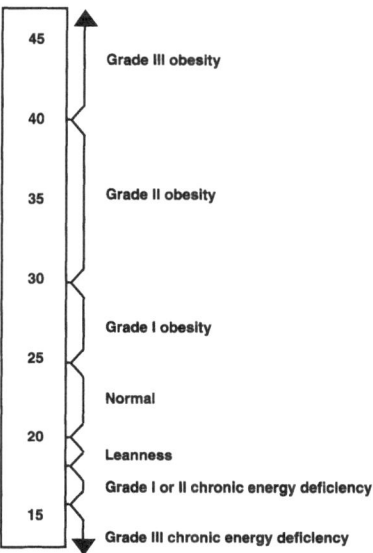

Fig. 1 Cut-off points of body mass index used for the determination of obesity, leanness and undernutrition.

grade II undernutrition was dependent upon the level of habitual energy intake (on energy expenditure) as compared to the basal metabolic rate. BMI below 16.0 (grade III) indicated severe energy deficiency whereas a BMI of 12 Kg/m² was considered as the absolute lower limit compatible with life (this would give a weight of 36.8 Kg for a height of 1.75 m). People suffering from chronic alcoholism often have BMI between 17 or 19 whereas young women suffering from anorexia nervosa have a BMI below 15 Kg/m². Some examples of low BMI occurring in certain nutritional conditions are shown in Table II.

It has to be stressed that the ideal body weight and the body mass index are arithmetically linked since both use weight for a given height as parameters. For example, a subject whose weight is 20% higher than his (her) ideal body weight by the Metropolitan Life Insurance Table[6] will have a body mass index of approximately 26 Kg/m².

The evaluations of the degree of overweight or underweight solely based on some ratio of weight to height (BMI, IBW) have limitations: none of them provide information upon the proportion of the 2 distinct body compartments, namely fat mass and fat-free mass. When assessing the nutritional status of a patient, it is of primary importance to obtain information on body composition since body weight

Table II. Low body mass index (BMI) reported in the literature (adapted from James et al,[3])

Subjects	Conditions	BMI (Kg/m²)
Indian adults	Severe protein-energy malnutrition	12
Young women	Anorexia nervosa	14-15
Indian labourers		16.6
American volunteers	Following semi-starvation experiments (24 weeks)	16.5
Columbian men	Malnourished	17.7

can be higher (or lower) than a "standard" value because of increased (or decreased) muscular mass rather than because of excess (or reduced) fat mass.

By using either the "ideal" body weight or BMI concept, one would incorrectly judge as obese individuals like weight lifters, body builders or American football players whose BMI can reach values over 30 Kg/m², but whose body fat percentage is within the normal range.[10] Obviously, this sample of athletes represents a low fraction of the total population so that the importance of incorrect diagnosis is minimal in epidemiological studies.

Figure 2 illustrates various combinations between excess or reduced fat-free mass combined with excess or reduced fat mass. For example, as compared with "normal" fat and "normal" fat-free mass individuals, patients suffering from chronic energy malnutrition or from anorexia nervosa have both a reduced fat and fat-free mass. It is also possible to encounter individuals who have low fat-free mass in conjunction with a "normal" fat mass, such as those chronically fed low protein diets of normal energy value: the protein intake below the protein requirement will lead to a negative nitrogen balance and explains the loss of fat-free mass.

Chronic overfeeding with high carbohydrate low protein diet would result in an elevated fat mass in conjunction with a "normal" or low fat-free mass.

Very lean individuals with low percentage body fat and apparently normal fat-free mass can be found among long distance runners. Obese individuals have by definition an excess relative body fat but it has to be stressed that their fat-free mass-when expressed in absolute value is also slightly elevated.

The degree of muscular hypertrophy, typically encountered in body builders, weight lifters or American football players results in an increased fat-free mass with a body fat which is "normal" or on the low side. Finally, one can also find isolated individuals (such as the Japanese wrestler called the "Sumo") with both drastically elevated fat mass and fat-free mass.

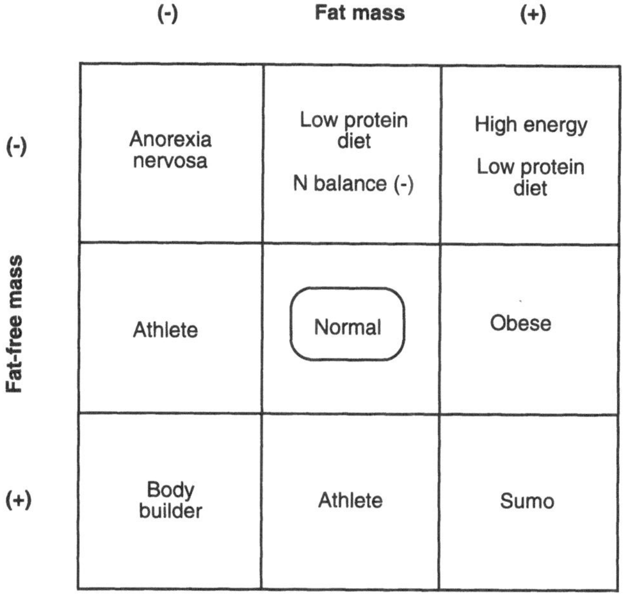

	(-) Fat mass (+)		
(-)	Anorexia nervosa	Low protein diet N balance (-)	High energy Low protein diet
	Athlete	Normal	Obese
(+)	Body builder	Athlete	Sumo

Fig. 2 Various combinations of low (-) to high (+) fat versus low to high fat-free mass.

(Fat-free mass, vertical axis)

Relationship between Body Mass Index (BMI) and Relative Body Fat

At a given weight and height (and hence BMI) the body composition is not the same for all individuals since there are interindividual variations in body fat. Due to the fact that the excess body weight of obese individuals is composed on average of 75% fat and 25% fat-free mass,[4] as previously mentioned, one can anticipate that an increased BMI results in an increased relative body fat. From a recent unpublished study performed in 759 normal individuals of the French part of Switzerland we have attempted to evaluate the relationship between percentage body fat (assessed by skinfolds) vs BMI. In adult men and women of various age intervals the correlation coefficient between these two variables was indeed highly significant ($p < 0.001$) but there was a lot of scatter in the relationship. As shown in figure 3, when the male and female data were combined, a wide range of percentage body fat for a given BMI was apparent: for example at a BMI of 20 Kg/m^2, the body fat ranged from 7 to 32%. Obviously this variability can be reduced when the men are studied separately from the women.

Thus for a given individual, the use of the BMI can lead to erroneous conclusions in terms of energy reserve. In contrast, for a large group of subjects (such as in epidemiological studies) the average BMI of the group will provide a useful

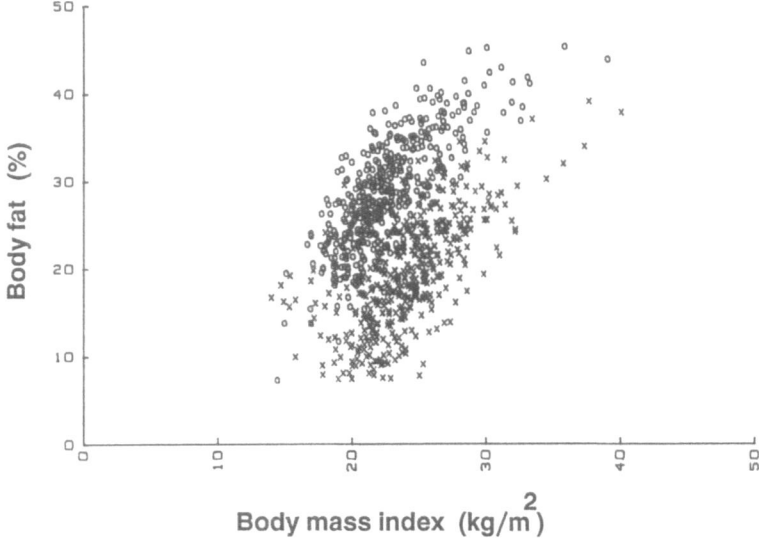

Fig. 3 Relationship between body mass index and relative body fat in 759 healthy individuals. O = female, X = male.

indication of excess body weight or obesity. Finally, due to the considerations mentioned above, one can anticipate that the relationship between BMI and percentage body fat is linear only within a certain range of BMI since the percentage body fat must eventually reach a plateau.

In addition to correctly evaluating the energy reserve of the organism, the importance of assessing body composition under clinical conditions is further emphasized by the fact that basal energy expenditure (BEE) is dependent upon body composition. It has been demonstrated[9] that the best predictor of BEE is fat-free mass (rather than body weight): BEE per unit fat-free mass is essentially the same for a man vs a woman or a young vs an elderly individual.

Lean Body Mass is not Equal to Fat-free Mass

It seems fundamental to recall the distinction between the concept of lean body mass vs fat-free mass (Fig. 4).

This distinction has to be made because the various methods available to assess body composition allow the determination of either one or the other parameters: for example, lean body mass can be determined by total body water or total body potassium whereas fat-free mass can be assessed by whole body densitometry. Lean body mass is defined as the total body weight minus the adipose tissue mass, whereas fat-free mass is defined as the total body weight minus the total body fat.

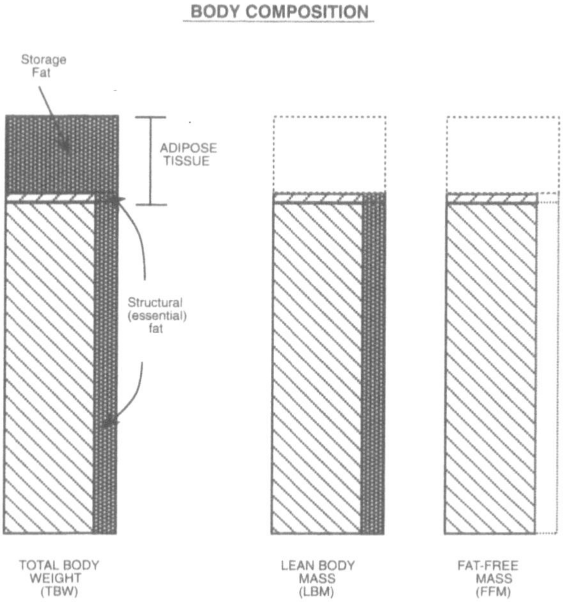

BODY COMPOSITION

Fig. 4 Distinction between the concept of lean body mass vs fat-free mass.

Therefore lean body mass comprises fat-free mass + the structural essential fat, which represents 3 to 5% of the body weight.

Stated briefly, fat-free mass does include neither essential structural fat nor storage fat. Although in the literature lean body mass and fat-free mass are two terms generally indistinctly used, it seems of importance to clarify the basic difference between these 2 concepts.

Clinical Methods Used to Assess Body Composition

The simplest method used to assess the fat mass compartments is the skinfold thickness. It is quick, easy to perform and inexpensive. It consists in measuring by means of a caliper (which exerts a constant pressure, typically 10 g/mm^3) the thickness of the skin + the subcutaneous fat layer at several sites. Classically 4 sites are used: the triceps, biceps, subscapular and suprailiac.[1] For normal or lean individuals, this measurement is easy to perform whereas for grossly obese subjects, some measurements are difficult to carry out, in particular the suprailiac site.

This method of body fat assessment is based upon two important assumptions: first the sites selected for the measurements must be representative of the entire topography of subcutaneous fat layer. Second, the total subcutaneous fat must reflect total body fat, i.e. internal (visceral) + peripheral subcutaneous fat. From the sum of the 4 skinfolds - each being measured in triplicate - an empirical equation such as that established by Durnin and Rahman permits the estimation of total body density. Since the average density of fat-free mass is 1.1 g per cubic centimetre, while the average density of fat mass is 0.9 g per cubic centimetre, this allows the calculation of the proportion of fat to fat-free mass in the body.

What are the reference values for percentage body fat in healthy men and women based on skinfolds for thickness? Figure 5 based on Schutz[10] presents the normal values in function of age for both sexes. It can be seen that in a normal young woman the percentage body fat can range from 20 to 30% whereas in man the normal range is 10 to 20% of body weight. Therefore, women have about 10% more fat than men. In our laboratory, the lower value we have obtained for a woman suffering from anorexia nervosa was 9% of body fat whereas the highest value measured in a grossly obese woman was 50% body fat. The upper and lower limits presented in figure 5 determine the limits of obesity and leanness respectively in function of age. These limits have been derived from the assumption that the amount of subcutaneous fat (as evidenced by skinfolds) should remain constant with aging. In fact this is usually not the case since most individuals from Western societies show a weight gain and a large increase in percentage body fat with aging! For example, a man with 22% body fat at the age of 20 will reach 30% body fat at the age of 60 despite constant skinfold readings. The higher relative body fat obtained at the age of 60 vs 20 can be explained by 2 factors: 1. total body fat to subcutaneous fat ratio may increase progressively with aging, due to a progressive increase in internal (visceral) fat storage and 2. aging is generally accompanied by progressive reduction in fat-free mass.

An important practical limitation of the skinfold method stems from the lack of access to all measurement sites in certain patients such as in critically ill or in burned patients. The use of one single skinfold thickness, i.e. triceps - as typically performed in routine clinical nutritional evaluations - cannot be used to safely predict total body fat and hence total energy reserve. Recently, new non-invasive methods have been developed to assess fat-free mass such as bioelectrical impedance.[5] It involves the use of a low electrical alternating current to measure electrical body resistance by means of electrodes located on the wrist and on the ankle of the body. This method is based upon the electrical properties of the human tissue: storage fat contains no water and electrolytes so that when a current is applied its bioelectrical resistance is high. On the other hand, the bioelectrical resistance of fat-

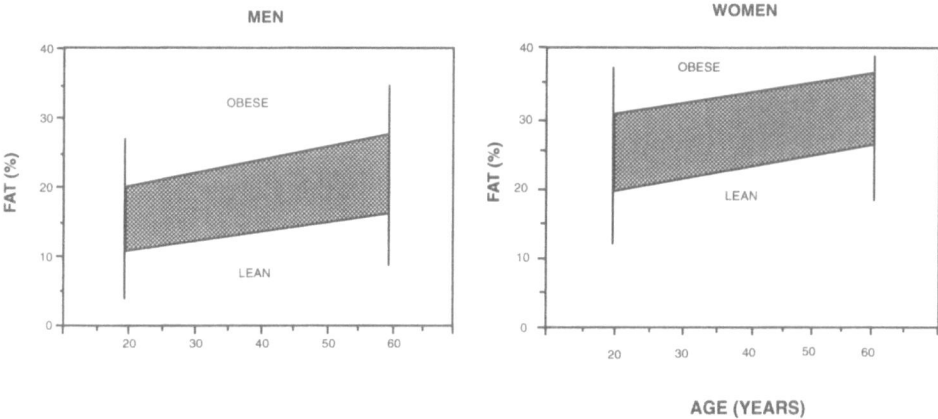

Fig. 5. Reference values for relative body fat in men and women of different ages.

free tissue is low since it contains virtually all the fluids and the body electrolytes. For a given individual, the measured bioelectrical resistance is thus inversely proportional to fat-free mass, i.e. when the resistance is low, the fat-free mass is high and conversely. For reasons which cannot be developed here, the length of the subject must also be taken into account and the results are expressed as the ratio between height squared divided by the measured resistance (H2/R).

Figure 6 shows the results of a cross-sectional study performed in the French part of Switzerland on 759 individuals (same sample as previously described). The correlation between the fat-free mass estimated from skinfolds vs electrical resistance expressed as H2/R - the latter being measured at a frequency of 50 kHz - was highly significant (r = 0.938, p<0.0001). It has to be stressed that part of the variability observed in the relationship is due to the variability of the skinfold method "per se" which cannot be considered as a reference standard. The potential advantages of assessing fat-free mass by using bioelectrical resistance (or imped-ance) are that it is non-invasive, non-expensive and the instrument is portable and can be easily used at the bedside of the patient.

In conclusion, the assessment of body composition is of primary importance for assessing the nutritional status of a patient. Excess body weight or deficient weight for a given height cannot provide any reliable estimates of the energy reserves in a given individual. Although numerous indirect methods have been developed over the years to determine body composition in clinical conditions, sophisticated time-consuming methods cannot be used. The two simple available methods are the skinfold thickness technique and the bioelectrical impedance. The latter needs to be further studied in patients presenting various degrees of fluid disturbances.

12

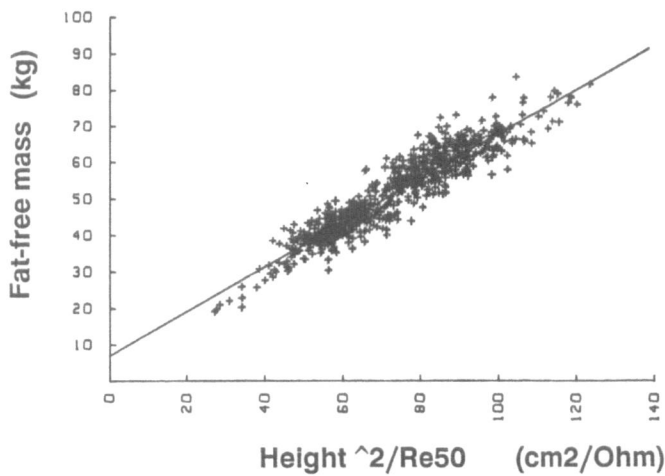

Height ^2/Re50   (cm2/Ohm)

Fig. 6 Regression line between fat-free mass on height squared divided by body resistance at 50 Khz (Re50) in 759 healthy men and women. r=0.938, p<0.0001. (Same data as figure 3).

References

1. Durnin J.V.G.A., Rahaman M.M.: Body fat assessed from total body density and its estimation from skinfold thickness: measurements on 481 men and women aged from 16 to 72. Br. J. Nutr. 1974; 32:77-97
2. Garrow J.S., Webster J.: Quetelet's index (W/H2) as a measure of fatness. Int. J. Obes. 1985; 9:147-153
3. James W.P.T., Ferro-Luzzi A., Waterlow J.C.: Definition of chronic energy deficiency in adults. Eur. J. Clin. Nutr. 1988; 42: 969-981
4. Jéquier E., Schutz Y.: Energy expenditure in obesity and diabetes. Diab./Metab. Rev. 1988; 4:583-593
5. Lukaski H.C., Johnson P.E., Bolonchuk W.W., Lykken G.I .: Assessment of fat-free mass using bioelectrical impedance measurements of the human body. Am. J. Clin. Nutr. 1985; 41:810-817
6. Metropolitan Life Insurance Company: New weight standards for men and women. Stat. Bull. 1959; 40:1-4
7. Metropolitan Life Insurance Company: Stat. Bull. 1983; 64:1-9
8. Quetelet LAJ: *Physique sociale* 2.Brussels, C Muquardt, 1869; pp 92
9. Ravussin E., Burnand B., Schutz Y., Jéquier E.: Twenty-four hour energy expenditure and resting metabolic rate in obese, moderately obese, and control subjects. Am. J. Clin. Nutr. 1982; 35:566-573
10. Schutz Y.: Consommation alimentaire et excès pondéral en Suisse. Méd. Hyg. 1984; 42:3864-3873

2. Serum and Immunological Parameters in the Assessment of Nutritional Status

P. DIONIGI, V. JEMOS, T.CEBRELLI

Department of Surgery, Istituto di Patologia Chirurgica, University of Pavia, Italy

Nutritional status is traditionally assessed by means of different indicators: clinical history and physical examination, anthropometry (weight, triceps skinfold midarm muscle circumference), body components (body fat, total body water, extracellular water, lean body mass), visceral proteins (albumin, transferrin, retinol binding protein, pre-albumin, fibronectin), immunoreactivity (delayed hypersensitivity response, complement components, lymphocyte count, immunoglobulins, leucocyte chemotaxis and phagocytosis) and functional tests (hand grip dynamometry, electrical stimulation).

All these parameters have been proposed and employed individually or in variable association:

1. to detect the presence of abnormalities in a population (nutritional epidemiology);
2. to identify patients at risk for complications or death related to malnutrition or to metabolic derangements (nutritional risk);
3. to evaluate the efficacy of the caloric and nitrogen supply (nutritional monitoring). We will briefly consider the use of visceral proteins and immunologic parameters for the first two points.

Nutritional Epidemiology

A few circulating proteins synthesized by the liver are considered good indicators of malnutrition; their blood levels are frequently impaired in hospitalized patients, and this could be related to a variety of nutritional and non nutritional factors. In chronic malnutrition low values of visceral proteins seem to be related more to a reduced hepatic synthesis than to an increased peripheral breakdown.[25] The relatively long half-life of albumin[35] and the considerable capacity of metabolic adaptation explain its low sensitivity in estimating acute depletion when compared to other serum proteins with rapid turnover rate like transferrin, retinol binding protein (RBP), and pre-albumin.[23] Infection, injury, myocardial infarction, malignant and acute diseases may affect circulating fluid volume, exchange with the extravascular tissue space, lymphatic return, catabolism, synthesis and losses of albumin and circulating proteins.[18] In fact the majority of these proteins act as positive or negative acute phase reactant, their blood concentration being increased or decreased as a consequence of different types of minor or major tissue injury. Nevertheless the overall acute-phase response correlates with undernutrition.[17, 33]

Malnutrition has been shown to be associated with reduced immunocompetence. Low levels of complement activity reflect a general reduction in protein synthesis, a consumption of some components or a shift towards a specific limited synthetic capacity. On the other hand, elevated complement fraction levels have been observed in inflammatory diseases and cancer.[15]

Multiple synergistic factors are likely to be involved in the etiology of depression of cellular immunity (nutrition, age, primary or acquired immunodeficiency, malignant disease, immunosuppressive therapy) so that some Authors believe that there is insufficient evidence supporting the use of delayed hypersensitivity response (DHR) as an indicator of nutritional status.[37] Nevertheless, in patients matched for age, sex, disease and nutritional status, DHR impairment appears to be caused mainly by aging and malnutrition.[16]

In developing countries, subjects with severe nutritional deficit present a depressed total lymphocyte count (TLC).[10] The same factors influencing DHR are known to interfere with this parameter. Moreover, bacterial infections can result in leucocytosis and lymphocytosis despite the presence of malnutrition.[27]

As suggested by Dempsey and Mullen,[11] to determine whether a parameter truly reflects a nutritional deficiency, different hospitalized populations should be studied: healthy inpatients (elective herniorrhaphy for instance), malnourished hospitalized subjects and wellnourished patients; this should enable control for non nutritional influences on the testing parameter under evaluation. We have compared 4 homogeneous groups of patients similar in age and sex distribution: a population with benign lesions not of the gastrointestinal tract undergoing minor surgery, subjects with advanced gastric cancer with and without clinical evidence of malnutrition (evaluated on the basis of history and physical examination) and a wellnourished group of patients with peptic ulcer. Table I shows that there is a good agreement between the clinical determination of nutritional status and the values of some anthropometric, visceral protein and immunologic parameters. In general, patients with gastric cancer present a higher degree of malnutrition when compared to subjects with a benign disease of the stomach. In this study, skin reactivity to

Table I. Epidemiological survey with nutritional indicators in hospitalized patients.

	Normal (n=21)	Peptic ulcer (n=24)	Gastric cancer Wellnourished (n=180)	Gastric cancer Malnourished (n=119)
Age	57±10	54±13	64±12	64±10
M/F	15/6	17/7	121/59	71/48
Stage III-IV (%)	-	-	72	84
UBW (%)	101±3	95±6 **	90±6 ***	87±10
Albumin (g%)	4.1±0.5	3.9±0.5 **	3.7±0.3 ***	2.8±0.3
Transferrin(mg%)	271±45	281±73	274±50 **	247±80
R.B.P. (mg%)	5.5±0.9	4.8±1.5 *	4.3±1.2	4.0±2.1
Ceruloplasmin (mg%)	38±4.2	45±10	43.3±10.3	46±12
Lymphocytes (mm³)	2797±731	3007±1068	2293±1153	2285±778
W.B.C.(mm³)	6714±1181	8213±2925	7462±3242	8524±3040
C3 (mg%)	94.4±17.3	116±26	117±24.5	123±37
DHR (%normoergic)	100	82 ***	38 *	25

*P<0.05; **P<0.01; ***P<0.001

antigenic challenge seems to be more influenced by the presence of the neoplasia than by malnutrition. On the contrary, percent usual body weight, albumin and transferrin levels are significantly lower in malnourished than in wellnourished patients with cancer. Therefore these parameters could be considered reliable indicators of the nutritional status in a population with cancer of the stomach. Moreover, it is possible to differentiate severe from mild malnutrition by increasing the cutoff point of the 3 parameters. Unfortunately few studies have appeared following Dempsey's suggestions, so that the results are often conflicting and the controversy continues about whether these variables are truly nutritional.

Nutritional Risk

Epidemiological surveys demonstrate that malnutrition affects a significant proportion of hospitalized patients,[5, 6, 14, 29, 32] but from the clinical point of view it seems to be more practical to define malnutrition on a prognostic basis as a nutritional deficit associated with an increased risk of adverse clinical events (complications, death). If this condition is recognized, patients at risk can be scheduled for an adequate nutritional support. After a nutritional parameter has been selected (for instance albumin), a cutoff point must be decided (less than 3.5 g%) and an end point indicated (death, postoperative septic complications).[19] The results must be analysed in terms of sensitivity, specificity, accuracy and overall predictive value, considering that variations of the cutoff point and of the end point influence sensitivity and specificity. The relationship between albumin levels and mortality is presented in Table II: sensitivity and specificity of the test are affected by the selection of the cutoff point of the parameter, the age and prevalence of malnutrition. The overall predictive value of the test in all the studies is quite elevated demonstrating that albumin can be considered a reliable indicator of mortality risk in different populations.

Dempsey and Mullen recently reviewed the available data of "single" nutritional assessment parameters related to a fatal outcome.[11] Table III summarizes the results of their analysis showing that death was correctly predicted in 81% of individuals

Table II. Prognostic value of albumin versus mortality.

	n	Alb (g/dl)	SE (%)	SP (%)	OPV (%)
Hay R.W.[21]	340	<2.5	95	18	83
Reinhardt G.F.[34]	2060	<2.5	43	97	89
		<3.0	67	90	89
		<3.4	85	78	79
Apelgren K.N.[3]	17	<2.5	100	82	88
Argawal N.[1]	80	<3.0	33	39	91

SE= sensitivity; SP=specificity; OPV=overall predictive value by albumin and in 82% by DHR. On the contrary, due to their low specificity, CHI (creatinine height index), transferrin and total lymphocyte count have a minor predictive value.

To increase the discriminating ability of the individual parameter for mortality and complications (mainly septic complications), multiparameter nutritional risk

Table III. Single nutritional assessment parameters as predictors of mortality.

	n	SE (%)	SP (%)	OPV (%)
Weight loss	46	86	69	72
CHI	225	65	58	60
Albumin	3019	69	82	81
Transferrin	229	77	39	48
Lymphocytes	857	76	62	63
Anergy	1738	52	86	82

SE= sensitivity; SP= specificity; OPV = overall predictive value (modified from Dempsey, 1987)

indices have been proposed. Buzby et al.[8] showed that surgical patients can be divided into 2 groups at low or high risk of mortality and postoperative complications by a prognostic nutritional index (PNI) based on albumin, transferrin, triceps skinfold and DHR.

Table IV demonstrates that varying the PNI cutoff point from 50 (high risk) to 40 (low risk), death is predicted with a greater sensitivity but less specificity.

In a surgical population Harvey et al.[20] developed a hospital prognostic index based on serum albumin, DHR, presence or absence of sepsis and cancer. The ability of this index to predict death in the tested population was not different from the previous ones (Table IV).

In a wide population admitted to the ICU, Seltzer [36] demonstrated that an abnormal albumin (<3.5 g/dl) is associated with a sixfold increase in mortality, and TLC less than $1500/mm^3$ with a fourfold increase in deaths; patients with both low albumin and low TLC presented a 20 times increase of mortality. Unfortunately the addition of TLC to albumin improves only the sensitivity of the test, but not its overall predictive value (Table IV).

Nazari assessed the nutritional status of a hospitalized surgical population by means of the contemporary multivariate and computerized analysis of anthropometric parameters, visceral proteins, muscle mass and immunological indicators.[31] Patients were divided by the computer into 4 similar clusters. Each cluster was then retrospectively analysed for patient distribution according to the prevalence of disease, incidence of postoperative complications, hospital stay and mortality. Most of the patients with benign diseases who underwent minor surgical procedures and had no postoperative complications were found in cluster 1, whereas

Table IV. Multiparameter nutritional indexes and risk of death.

	n	SE (%)	SP (%)	OPV (%)
Buzby G.P.[8]				
PNI >50	145	86	69	72
PNI >40	145	93	44	51
Harvey K.B.[20]				
HPI	282	74	66	72
Seltzer M.H.[36]				
Alb <3.5 g%	120	43	76	70
TLC<1500/mm^3	120	83	48	54
Alb and TLC	46	71	70	71

SE= sensitivity; SP= specificity, OPV = overall predictive value

patients with progressive major diseases and worse prognosis were in clusters 2-3-4. Nevertheless, the clinical usefulness of this interesting technique was not fully demonstrated, as is reported afterwards. Other studies employing multiple variable analysis of nutritional parameters have recently confirmed that only serum albumin is less affected by age, correlates significantly with most of the measured nutritional indices and remains the best predictor of the probability of death.[1, 9, 28]

In conclusion, many mono- and multi-parameter indices composed of visceral proteins and/or immune reactants have been proposed to detect surgical patients with nutritional impairment at risk of postoperative complications. Only a few Authors[8, 31] validated their index with a prospective study, but no one has compared different indices within the same population.

In 1985 thirty-five Italian hospitals were involved in a multicentric trial to compare the clinical value of different methods of nutritional assessment in order to recognize patients at risk of postoperative infectious complications. A total of 548 patients was examined, in 122 cases data were insufficient. A standard nutritional assessment was performed before surgery using anthropometric, visceral protein and immunological parameters expressed as % of the standard reference value. A total of 140 over 426 patients presented one or more postoperative infectious episodes (urinary, respiratory, wound infection, intra-abdominal abscess, thrombophlebitis). Albumin, transferrin and % usual body weight analysed as single parameters were able to stratify septic patients proportionally to the degree of malnutrition, but sensitivity, specificity, accuracy and overall predicting value of each of them was low. All the anthropometric parameters, visceral proteins and immune responses were then expressed as a mean of the % values. Albumin and TLC were also associated in a index according to Seltzer,[36] the PNI of Buzby and Mullen was derived[30] and patients were attributed to one of the 4 clusters according to Nazari's method.[31] Table V demonstrates that each index has a relatively low overall predictive value for postoperative infections. PNI sensitivity, specificity, accuracy and predictive value are very close to those of the visceral protein suggesting that albumin and transferrin play a major role in determining the final result.

Similarly immunological parameters (more than visceral proteins), seem to influence the cluster analysis predictive value. In summary none of the single or multiple nutritional parameters used in this survey presents a sufficient accuracy to

Table V. Italian multicentric trial 1985.[24] Risk of postoperative septic complications (n=426).

	SE (%)	SP (%)	OPV (%)
Anthropometrics	61	40	46
Visceral proteins	58	60	59
Immunological parameters	50	77	24
Seltzer	62	45	50
PNI	52	69	64
Cluster analysis	77	49	59

identify patients at risk of postoperative septic complications. This could be due to inadequate reference standards or to the heterogeneity of the population studied, but it is evident that non nutritional factors can influence the postoperative risk of sepsis, like underlying disease,[31] duration of anaesthesia,[14] catheters, bacterial contamination of the operating field,[2] pre-operative blood losses and transfusions,[26] and therapeutic procedures (antibiotics).

In the last few years an intensive statistical analysis was carried out to clarify the relationship between nutritional indices and postoperative complications. The purpose of the studies was to minimize the interference of non nutritional factors in order to recognize patients at risk that could benefit from a nutritional treatment. Detsky et al. have recently proposed a useful method to investigate the relative predictive accuracy of nutritional assessment techniques.[12] After the sensitivity and specificity of a test are determined, by plotting positive versus negative predicting values, the optimal cutoff point of that parameter can be defined. Since at a given prevalence of malnutrition within the population studied the sensitivity and specificity of the test are defined by the selected cutoff point, different nutritional assessment parameters become comparable. The results of these studies have been sometimes surprising, demonstrating a better predictive value of clinical parameters (subjective global assessment based on history and physical examination), than visceral proteins, immunological parameters, PNI or other indices. The same group of investigators has applied decision analysis to compare 3 strategies for reducing the incidence of nutrition-associated complications in patients undergoing major gastrointestinal surgery: treat none, treat all, treat selectively.[12] Their results indicate that the optimal strategy for the lowest complication rate is dependent on 4 factors: prevalence of nutrition-associated complications, effectiveness of parenteral nutritional support, iatrogenic complication rate and the characteristics of the test, such as sensitivity, specificity and predictive value.

In conclusion, biochemical indicators and immunological parameters have been utilized in epidemiological nutritional surveys to identify malnourished populations. The same parameters demonstrate a limited value when applied to a single patient, especially when a borderline malnutrition is present. The evaluation of the risk of developing nutrition-related complications is a more interesting field of investigation for clinicians, but results may be affected by non nutritional factors. At present, statistical methods for the development and validation of nutritional assessment techniques are available, but we need perspective multicentric trials in homogeneous hospitalized patients.

References

1. Agarwal N., Acevedo F., Leighton L. S. et al.: Predictive ability of variables for mortality in elderly people. Am. J. Clin. Nutr. 1988; 48: 1173-8

2. Altemeier W.A., Burke J.F., Pruitt B.A. Jr, Sandusky W.: Manual on control of infections in surgical patients. Committee on control of surgical infections of the community on pre- and postoperative care. American College of Surgeons, Philadelphia, JB Lippincott Company 1976

3. Apelgren K.N., Rombeau J.L., Twomey P.L .et al.: Comparison of nutritional indices and outcome in critically ill patients. Crit. Care. Med. 1982; 10: 305-307

4. Baker J.P., Detsky A.S., Wesson D.E. et al.: A comparison of clinical judgment and objective measurement. N. Engl. J. Med. 1982; 306: 969-972

5. Bistrian B.R., Blackburn G.L., Vitale J., Cochran D., Naylor J.: Prevalence of malnutrition in general medical patients. JAMA 1976; 235, 1567-70

6. Bozzetti F.: Nutritional assessment from the perspective of a clinician. J. Parent. Ent. Nutrition 1987; 11 (Suppl): 115 S - 121 S

7. Brenner U., Muller J.M., Keller H.W., Walter M.: Nutritional assessment in surgical planning. Clin. Nutr. 1988; 7:225-229

8. Buzby G.P., Muller J.L., Matthews D.C., Hobbs C.L., Rosato E.F.: Prognostic nutritional index in gastrointestinal surgery. Am. J. Surg. 1980; 139:160-167

9. Casey J., Flinn W.R., Jao J.S.T., Fahey V., Pawlowsky J., Bergan J.J.: Correlation of immune and nutritional status with wound complications in patients undergoing vascular operations. Surgery 1983; 93:822-827

10. Chandra R.K.: Rosette forming B-lymphocytes and cell mediated immunity in malnutrition. Br. Med. J. 1974; 3:608-9

11. Dempsey D.T., Mullen J.L.: Prognostic value of nutritional indices. J. Parent. Ent. Nutrition. 1987; 11 (Suppl): 109 - 114 S

12. Detsky A.S., Baker J.P., Mendelson R.A., Wolman S.L., Wesson D.E., Jeejeebhoy K.N.: Evaluating the accuracy of nutritional assessment techniques applied to hospitalized patients: methodology and comparisons J. Parent. Ent. Nutrition 1984; 8:153-159

13. Detsky A.S., Baker J.P., O'Rourke K. et al.: Predicting nutrition associated complications for patients undergoing gastrointestinal surgery. J. Parent. Ent. Nutrition 1987; 11:440-446

14. Dionigi P., Dionigi R., Nazari S. et al.: Nutritional and immunological evaluations in cancer patients. Relationship to surgical infections. J. Parent. Ent. Nutrition 1980; 4: 351-356

15. Dionigi P., Nazari S., Bonoldi A.P., Cividini F., Olezza S., Dionigi R.: Nutritional assessment and surgical infections in patients with gastric cancer or peptic ulcer. J. Parent. Ent. Nutrition 1982; 6:128-133

16. Dominioni L., Dionigi R., Dionigi P. et al.: Evaluation of possible causes of delayed hypersensitivity impairment in cancer patients. J. Parent.Ent. Nutrition 1981; 5: 300-306

17. Dominioni L. Dionigi R., Jemos V.: The acute phase response of plasma proteins in surgical patients. In: Wesdorp R.I.C., Soeters P.B. (Eds), *Clinical Nutrition* 81. London,Churchill Livingstone 1982; 239-259

18. Fleck A., Raineg G., Hawker F. et al.: Increased vascular permeability: a major cause of hypoalbuminemia in disease and injury. Lancet 1985; i:781-783

19. Habicht J.P., Meyers L.D., Brownie C.: Indicators for identifying and counting the improperly nourished. Am. J. Clin. Nutr. 1982; 35:1241-1254

20. Harvey K.B., Moldawer L.L., Bistrian B.R., Blackburn G.L.: Biological measures for the formulation of a hospital prognostic index. Am. J. Clin. Nutr. 1981; 34:2013-2022

21. Hay R.W., Whitehead R.G., Spicer C.C.: Serum albumin as a prognostic indicator in oedematous malnutrition. Lancet 1975; ii:427-429

22. Inagaki J., Rodriguez V., Bodey G.P.: Causes of death in cancer patients. Cancer 1974; 33:568-573

23. Ingenbleek Y., Van den Schrieck H.G., de Nayer P., De Visscher M.: Albumin, transferrin and

the thyroxine-binding protein (TBPA-RBP) complex in assessment of malnutrition. Clin. Chim. Acta 1975; 63:61-67

24. Italian Study Group for nutritional assessment. Multicentric trial on nutritional assessment methods. Identification of high risk surgical patients. Abstract book ESPEN Congress, Munich 9-11 September, 1985; 180

25. James W.P.T.: Albumin metabolism: effect of the nutritional status and the dietary protein intake. J. Clin. Invest. 1968; 47:1958-1972

26. Jubert A.V., Lee E.T., Hersh E.M., McBride C.M.: Effects of surgery, anesthesia and intraoperative blood loss on immunocompetence. J.Surg. Res. 1973; 15:399-403

27. Lewis R.T., Klein H.: Risk factors in postoperative sepsis. Significance of preoperative lymphocytopenia. J. Surg. Res. 1979; 26:365-371

28. Mitchell C.O., Lipschitz D.A.: The effect of age and sex on the routinely used measurements to assess the nutritional status. Am. J .Clin. Nutr. 1982; 36:340-349

29. Mullen J.L., Gertner M.H., Buzby G.P., Goodhart G.L., Rosato E.F.: Implications of malnutrition in the surgical patient. Arch .Surg. 1979; 114:121-125

30. Mullen J.L., Buzby G.P., Matthews D.C., Smale B.F., Rosato E.F.: Reduction of operative morbidity and mortality by combined preoperative and postoperative nutritional support. Ann. Surg. 1980; 192:604-613

31. Nazari S., Comincioli V., Dionigi R. et al.: Cluster analysis of nutritional and immunological indicators for the identification of high risk surgical patients. J. Parent. Nutrition 1981; 5:307-316

32. Nixon D.W., Heymsfield S.B., Cohen A.E. et al.: Protein calorie undernutrition in hospitalized cancer patients. Am. J.Med. 1980; 68: 683-690

33. Pomposelli J.J., Flores E.A., Bistrian B.R.: Role of biochemical mediators in clinical nutrition and surgical metabolism. J. Parent. Ent.Nutrition 1988; 12:212-218

34. Reinhardt G.F., Myscofski J.W., Wilkens D.B., Dobrin P.B., Mangan J.E., Stannard R.T.: Incidence and mortality of hypoalbuminemic patients in hospitalized veterans. J. Parent. Ent. Nutrition 1980; 4:357-359

35. Rothschild M.A.,Oratz M., Schreiber S.S.: Albumin synthesis. N. Engl. J. Med. 1972; 286:748-757

36. Seltzer M.H., Fletcher H.S., Slocum B.A., Engler P.E.: Instant nutritional assessment in the intensive care unit. J. Parent. Ent. Nutrition 1981; 5:70-72

37. Twomey P., Ziegler D., Rombeau J.: Utility of skin testing in nutritional assessment: a critical review. J. Parent. Ent. Nutrition 1982; 6:50-58

3. Energy Balance

Y. SCHUTZ

Institute of Physiology, Faculty of Medicine, Lausanne University, Switzerland

Classically, the concept of energy balance has been presented by a simple equation:

Energy balance = energy in - energy out

In energy equilibrium (zero energy balance), energy in (E_{in}) equals energy out (E_{out}), the net energy storage is zero and the total body energy is maintained constant. In positive energy balance, E_{in} is greater than E_{out} and the subject is storing energy in form of fat, protein and glycogen. In negative energy balance, E_{out} is greater than E_{in} and the subject is mobilizing energy from endogenous stores in the form of glycogen, protein and fat. It should be pointed out that E_{in} represents the energy physiologically available to the organism, the so-called "metabolizable" energy intake. The metabolizable energy is determined from the heat of combustion of food (i.e. the "gross" energy) minus the heat of combustion of urinary plus fecal energy losses.[12]

This apparently simple equation hides a number of potential problems: First, in free-living ambulatory conditions, there is a day-to-day variability of energy intake as well as energy expenditure, however not necessarily in the same direction. The within-subject variability of energy intake is known to be substantially greater than the within-subject variability of energy expenditure (expressed as the coefficient of variation). This implies that a single day measurement of energy intake or/and energy expenditure is clearly insufficient to obtain representative results of both variables.

These limitations do not necessarily apply to the hospitalized patients since the energy intake is generally imposed by the clinical situation and the physical inactivity of a patient confined to bed damps the day-to-day variations in E_{out}.

Secondly, except in drastic underfeeding or overfeeding conditions, the energy balance represents the difference between two numbers of the same order of magnitude so that the energy balance results will be uncertain. In other words, a relatively small error on E_{in} or E_{out} (or both) will lead to a large error in energy balance calculation.

Why do we Need to Assess Energy Balance in Hospitalized Patients?

The following reasons can justify the assessment of energy balance in the hospital:
1. The changes in body weight over a short period of time (i.e. a few days) represent a poor indicator of the changes in energy balance since fluid retention or mobilization can represent a confounding factor. In contrast, long-term body weight and body composition changes provide a suitable indicator of the trend in energy balance.
2. At a constant nitrogen (protein) intake, the level of energy intake - and thus energy balance - has a major impact on nitrogen balance. In healthy subjects, it is well known that the maintenance of nitrogen balance largely depends upon an adequate energy supply.
3. During the nutritional "rehabilitation" following illness or trauma, overfeeding should be discouraged in order to avoid excess fat deposition.

This report will focus on the E_{out} component of the energy balance only.

Total Energy Expenditure (E_{out})

E_{out} represents the total heat production of the subject, i.e. the total energy expenditure. Indirect calorimeter has been the method of choice to assess total energy expenditure in both healthy and sick individuals. Direct calometry - which assesses total heat losses - has been used primarily for thermoregulation studies. The combination of direct and indirect calorimetry allows calculation of the heat balance and hence the heat storage.[8]

When a subject maintains his core temperature constant (thermoneutral environment), total heat production assessed by indirect calorimetry is equal to total heat losses obtained by direct calorimetry and the heat storage is equal to zero. Due to the high complexity and high maintenance cost of direct calorimetry as well as the lack of direct access to the patient, the utilization of this technique in clinical conditions should be discouraged.

Indirect calorimetry is based on the measurement of oxygen consumption ($\dot{V}O_2$), carbon dioxide production ($\dot{V}CO_2$) and total urinary nitrogen excretion. In the past, face masks or mouth pieces have been utilized to collect expired air over a limited period of time. The use of the ventilated canopy system - proposed by

Kinney in the seventies - has the advantage of not disturbing the breathing pattern of the subject and allows continuous energy expenditure measurements over a prolonged period of time in bed-ridden patients. It should be pointed out that the only technique available today to assess 24 h energy expenditure continuously is the respiration chamber.[7] Due to the lack of direct access to the patient however, this technique cannot be applied to measure 24 h energy expenditure in hospitalized patients. Since in bed-ridden patients the resting energy expenditure may be close to the total energy expenditure, extrapolation of short repeated indirect calorimetric measurements over a 24 h period may be possible in order to obtain an estimate of 24 h energy expenditure.

This is particularly true since the circadian rhythm observed in healthy free-living individuals may play a minor role in patients confined to bed. In a recent study performed in healthy subjects, we have observed that the circadian energy expenditure oscillations of resting energy expenditure can be abolished by continuous enteral nutrient administration and constant rate of physical activity over 24 h (Schutz et al, unpublished study). The stability of day-time resting energy expenditure in healthy subjects maintained with continuous enteral nutrition had already been previously demonstrated.[14] According to Chikenji et al.[1] reliable estimates of 24 h resting energy expenditure in the clinical situation may be derived from 3 to 5 accurate measurements of gas exchange of 20 to 30 min. duration, evenly spaced throughout the 24 h period.

The assessment of 24 h energy expenditure by means of a respiration chamber has the advantage of being very accurate, but due to the confined situation, the rate of energy expenditure obtained may not be necessarily equivalent to that of free-living conditions. The accurate assessment of 24 h energy expenditure in the free-living situation has plagued physiologists and nutritionists for several decades. The classical techniques available have been previously reviewed[11] and will not be discussed here in detail. The most recent technique developed in humans is the doubly-labelled water method ($^2H_2^{18}O$), which can assess the rate of VCO_2 over a 10 to 15 day period. The rate of carbon dioxide production is converted into energy expenditure via the energy equivalent of VCO_2. The potential limitations of this elegant method should be kept in mind.[8] This essentially includes the cost of isotope, the cost of analysis by mass spectrometry, the various correction factors and the uncertainty of converting VCO_2 into energy expenditure when the RQ is unknown. The respective advantages and shortcomings of the respiration chamber vs the $^2H_2^{18}O$ method have been previously reviewed.[9]

The Different Components of Total Energy Expenditure

It has been customary to partition the total energy expenditure of a healthy individual into 3 components:

1. basal energy expenditure;
2. diet induced thermogenesis;
3. physical activity expenditure.

The basal energy expenditure represents the energy expended in basal standard-ized conditions, i.e. post-absorptive state (10-12h after the last meal), thermoneutral conditions, immobility but awake, subject without anxiety and fever. These strict basal conditions are difficult to obtain in a hospitalized patient. When these conditions are not met, we usually talk about resting (rather than basal) energy expenditure. It should be recalled that resting energy expenditure is largely influenced by the level of nutrient intake and the type of nutrient administered. Diet-induced thermogenesis (DIT) refers to the net increase in energy expenditure over the basal value. In sick patients, there is one additional component of the total energy expenditure which must be considered, i.e. the effect of illness and fever *per se*, which represent a factor of hypermetabolism. For example, injury, sepsis and burns produce a variable increase in resting energy expenditure (see Kinney, present issue). The extent to which COPD leads to an increased basal or resting energy expenditure is discussed separately (see Fitting, present issue).

In order to establish the magnitude of hypermetabolism in different pathological situations, the measured basal (resting) energy expenditure of the patient must be compared to a reference value. Ideally, a matched healthy control group should be studied and compared to the sick group. Unfortunately, this approach is rarely used in the clinical literature. Alternatively, the basal energy expenditure can be calculated from the anthropometric characteristics of the patient. From early this century, many formulae have been developed to predict basal energy expenditure. The formula which is most often used in clinical literature is that of Harris and Benedict, published in 1919.[4] The variables used to predict basal energy expenditure include sex, weight, height and age. Alternatively, the table of Fleisch[3] can be used but the results are given per square meter (m²) body surface area, the latter being calculated from the Du Bois equation using height and weight. More recently, the FAO/WHO/UNU expert committee[2] has suggested predicting basal energy ex-penditure of healthy men and women, classified into different age categories, from a series of simple linear regression equations (Table I).

If one makes a compilation of the different basal energy expenditure standards developed early in the century, one should realize that there are certain disparities between them. As shown in figure 1, when the different standards are displayed graphically in function of age, the variability between standards is very apparent. In the young, the maximum difference in prediction is in the order of 2 Kcal/m² body surface area per hour (0.06 Kcal/min. for an individual of 1.75 m²) whereas in elderly individuals, this difference is about twice as high. In practice, this indicates that the net hypermetabolism calculated on the basis of a predictive formula will depend upon which formula is used; for example, a sick elderly woman of 75 years

Table I. Equations for predicting basal energy expenditure (BEE) from body weight[2] (FAO/WHO/UNU 1985).

Age range	Women
18-30 y	BEE (kcal/d)= 496 + 14.7 x weight (kg)
30-60 y	BEE (kcal/d)= 829 + 8.7 x weight (kg)
> 60 y	BEE (kcal/d)= 596 + 10.5 x weight (kg)

Age range	Men
18-30 y	BEE (kcal/d)= 679 + 15.3 x weight (kg)
30-60 y	BEE (kcal/d)= 879 + 11.6 x weight (kg)
> 60 y	BEE (kcal/d)= 487 + 13.5 x weight (kg)

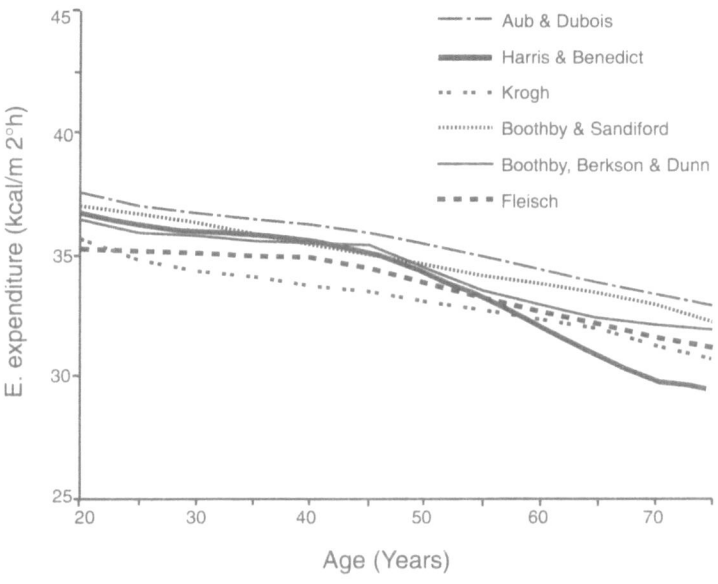

Fig. 1. Comparison between various basal energy expenditure (EE) standards developed by various investigators from the early century (for details, see Fleisch[3], 1951). Note that the progressive decrease in basal EE induced by aging varies between standards.

(whose resting energy expenditure is found to be elevated) will be considered relatively more hypermetabolic by Harris and Benedict[4] than by Fleisch[3] standards.

Energy Balance versus Nutrient Balance

Accurate nutrient balance determination can be best performed under laboratory conditions since the nutrient intake can be strictly controlled and the nutrient oxidation can be calculated by means of a whole body respiration chamber.[5,7]

The carbohydrate and fat balances represent the difference between the "metabolizable" carbohydrate and fat intakes minus the total carbohydrate and fat oxidations respectively, the latter being calculated by indirect calorimetry. The protein balance is calculated from the nitrogen balance, which is equal to the difference between total "metabolizable" nitrogen intake minus total urinary nitrogen excretion with an allowance for miscellaneous nitrogen losses (skin, etc.). In anabolic state when the energy equivalents of glycogen, fat and protein retention are known, the energy storage can also be calculated using appropriate factors.[12]

As shown in figure 2, energy balance, nutrient balance and acute changes in body composition represent 3 complementary measurements which can independently assess the net energy stored or mobilized. However, the nutrient balance is the only method available which allows the assessment of short-term changes in nutrient storage or mobilization since the methods available today for assessing body composition are not sufficiently precise to assess small changes in macronutrient storage (see Frascarolo et al. present issue). The accuracy with which the net changes in body composition can be assessed by means of the nutrient balance

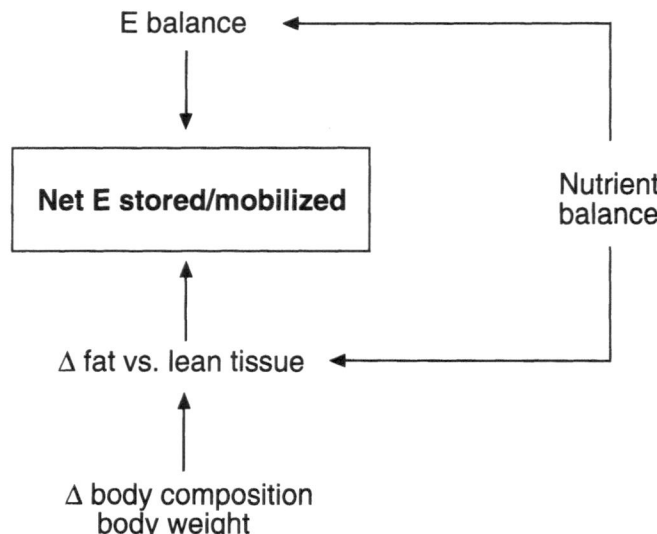

Fig. 2. Various complementary approaches to estimate the net energy storage or mobilization.

approach will depend upon both the accuracy of metabolizable nutrient intake and total nutrient oxidation. Recently, we have attempted to assess the nutrient balance in patients operated for oropharyngeal cancer during the first 10 days following surgery.[14] Whereas the energy and nutrient balance were close to equilibrium initially (day 5 post-surgery) energy, carbohydrate, fat and nitrogen balances were largely positive at the end of the study period (day 9 post-surgery). However, based on the body weight change over the 10-day study period, one would have erroneously judged the patients in negative energy balance since on average they were losing weight. This again emphasizes that body weight change does not provide an accurate estimate of change in energy storage in hospitalized patients.

Adaptation to Negative Energy Balance

Negative energy balance is a common feature in clinical conditions and this leads to progressive loss of body energy in the form of glycogen fat and protein. The negative energy balance can originate either from an increased resting (basal) energy expenditure and/or too low a level of metabolizable energy intake. Several early studies have been performed in volunteers who were subjected to starvation or semi-starvation diets. These experimental studies, as well as the investigations performed in overweight subjects who have lost weight[9] have clearly indicated that when an individual is losing weight his energy expenditure decreases. Figure 3 shows the relationship between the relative weight loss and the decrease in resting VO_2 in the studies performed under controlled conditions in normal (neither obese nor undernourished) individuals subjected to spontaneously or experimental restrictive diets for several weeks or months. The results shown have been based on those presented in tabular form by James[6] in a recent review. The basic question is to know whether the magnitude of reduction in energy expenditure during negative energy balance is equivalent to the reduction in body weight. In other words, does an individual who looses 20% of his (her) body weight reduce also his (her) energy expenditure by 20%?

In figure 3, the line of identity indicates the concordance between relative weight loss (expressed in percentage of initial weight) and relative decrease in basal oxygen consumption (expressed also as a percentage of initial value). Firstly, it can be seen that the greater the relative weight loss the greater the relative reduction in basal VO_2. Secondly, despite the heterogeneity of these data (number of subjects, type of diet), it is apparent that when the subject is measured while dieting, the relative decrease in basal VO_2 is, in most situations, greater than the decrease in relative weight loss. This indicates that after several weeks or months of negative energy balance, some metabolic adaptations have occurred via energy sparing mechanisms. It should be recalled that the resting energy expenditure is related to the fat-free tissue mass as well as to the plane of nutrition: the reduced food intake

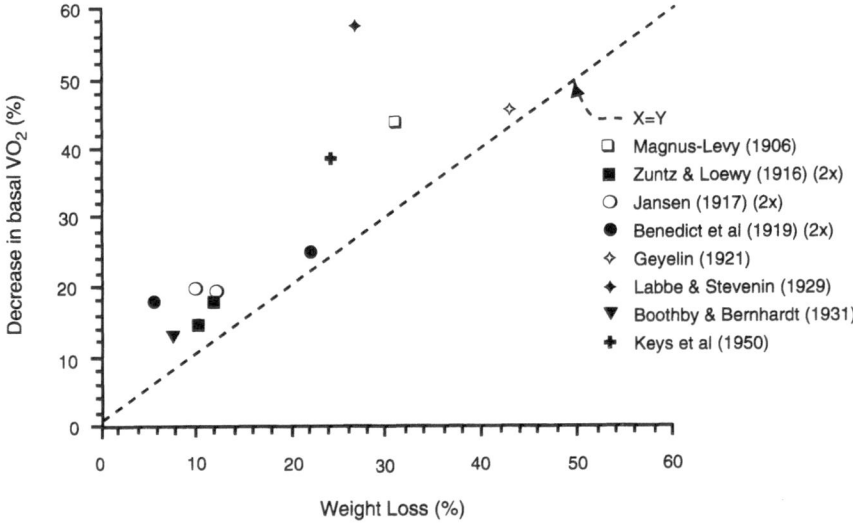

Fig. 3. Relative reduction in basal oxygen consumption (VO_2) induced by weight loss in various experimental studies in man. The dotted line represents the line of identity. Only the studies in which spontaneously restricted or experimental diets in non-obese normally nourished men and women were selected from a table published by James[6] (1988). Each point represents one study performed in a variable number of individuals.

per se leading to a fall in diet-induced thermogenesis as well as the reduction in fat-free mass induced by underfeeding could explain the overproportional fall in basal VO_2. A simplified diagram showing the mechanism by which the total energy expenditure decreases in response to chronic negative energy balance is presented in Fig. 4. The decrease in total energy expenditure counteracts the negative energy balance so that after a period of time a new energy balance equilibrium can eventually be reached.

Finally, it should be noted that the metabolic adaptations described above represent a confounding factor in hypermetabolic patients under chronic negative energy balance since two opposite phenomena are occurring simultaneously, the low energy intake decreasing and the disease state increasing energy expenditure. This conjunction of opposite effects may explain why some patients have paradoxically "normal" basal (resting) energy expenditure values in the presence of disease which is known to engender significant hypermetabolism.

To summarize, in the context of the daily clinical routine, it seems difficult to assess energy balance in hospitalized patients. For the clinician, knowledge of the dynamic day-to-day changes in energy balance appears to be more informative than the absolute energy balance calculation since the latter can be easily con-

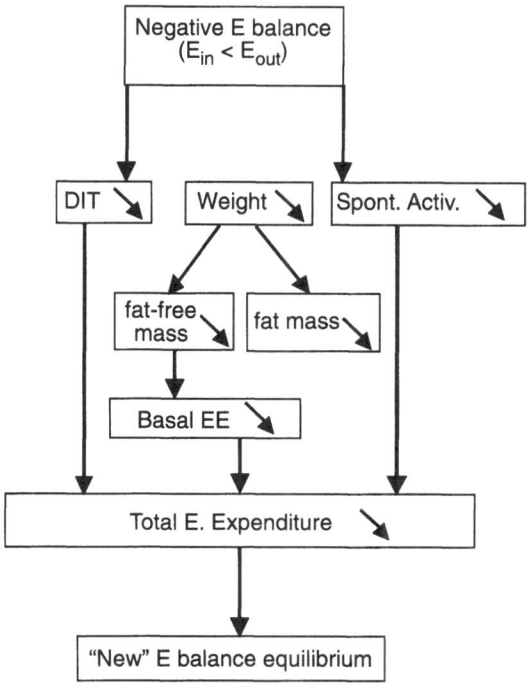

Fig. 4 Simplified schematic diagram explaining the mechanism by which total energy expenditure decreases with negative energy balance. DIT = diet induced thermogenesis. EE = energy expenditure.

founded by systematic errors on both energy intake or/and energy expenditure. The empirical formulae developed from the early century for predicting basal energy expenditure[3,4] can be used as guidelines for estimating the normal basal energy expenditure from the anthropometric characteristics of the patient but the error of prediction for a given individual may be substantial.

Calculation of the degree of hypermetabolism in a given patient will be uncertain since it will depend- particularly in elderly patients - upon which formula is used to predict the normal value. The only alternative to this uncertainty is to measure the energy expenditure of a matched control group in a state of energy balance which is similar to the group of patients under investigation.

References

1. Chikenji T., Elwyn D.H., Gil K.M., et al.: Effects of increasing glucose intake on nitrogen balance and energy expenditure in malnourished adult patients receiving parenteral nutrition. Clin. Sci.

1987; 72: 489-501 as cited by Bursztein S., Elwyn D.H., Askanazi J. and Kinney J.M. (Eds) In: *Energy Metabolism, Indirect Calorimetry, and Nutrition.* Baltimore, Williams & Wilkins 1989

2. FAO/WHO/UNU: *Expert consultation on energy and protein requirements.* Technical Report Series 724 WHO, Geneva 1985

3. Fleisch A.: Le metabolisme basal standard et sa determination au moyen du "Metabocalculator". Helv. Med. Acta, 1951; 18:23-44

4. Harris J.A., Benedict F.G.: *A biometric study of basal metabolism in man.* Washington Carnegie Inst. 1919

5. Hurni M., Burnand B., Pittet P., Jéquier E: Metabolic effects of a mixed and a high-carbohydrate low-fat diet in man, measured over 24h in a respiration chamber. Brit J. Nutr. 1981; 47: 33-43

6. James W.P.T.: Research relating to energy adaptation in man. In: Schürch B., Scrimshaw N.S. (Eds) *Chronic Energy Deficiency: Consequences and Related Issues.* Lausanne, Fondation Nestlé, 1988; 7-36.

7. Jéquier E., Schutz Y.: Long-term measurements of energy expenditure in humans using a respiration chamber. Am. J. Clin. Nutr. 1983; 38: 989-98

8. Jéquier E., Acheson K., Schutz Y.: Assessment of energy expenditure and fuel utilization in man. Ann. Rev. Nutr. 1987; 7: 187-208

9. Jéquier E., Schutz Y.: Classical respirometry and the doubly-labelled water ($^2H_2{}^{18}O$) method: appropriate applications of the individual or combined techniques. Proc. Nutr. Soc. 1988; 47: 219-225

10. Jéquier E., Schutz Y.: Energy expenditure in obesity and diabetes. Diab. Metab. Rev. 1988; 4: 583-593

11. Schutz Y.: Use of non-calorimetric techniques to assess energy expenditure in man. In: Björntorp P., Cairella M., Howard An (Eds): *Recent Advances in Obesity Research:* III. London, John Libbey 1981; 153-158

12. Schutz Y.: Terminology, factors and constants in studies on energy metabolism of humans. In: *Human energy metabolism: Physical activity and energy expenditure measurements in epidemiological research based upon direct and indirect calorimetry.* Report of an EC Workshop, Euro Nut Wageningen, The Netherlands 1984; 153-168.

13. Zurlo F., Schutz Y., Frascarolo P., Enzi G., Deriaz O., Jéquier E.: Variability of resting energy expenditure in healthy volunteers during fasting and continuous enteral feeding. Crit. Care Med. 1986; 14: 535-538.

14. Zurlo F., Schutz Y., Pichard C., et al.: What energy level is required to avoid nutrient depletion after surgery in oropharyngeal cancer? ORL 1988; 50: 236-245

Pathophysiology of Malnutrition in COPD

4. Epidemiology of Malnutrition in Chronic Obstructive Pulmonary Disease

S. R. BRAUN

Pulmonary, Critical Care and Environmental Medicine, University of Missouri, Columbia, USA

Introduction

Physicians have observed for many years that markers of nutritional depletion were present in individuals with chronic obstructive pulmonary disease (COPD).

In 1914 Dr. H.A. Hare stated, in regard to patients with emphysema, "Nutrition suffers and the patient grows weak and emaciated".[1]

Attempts to describe different COPD classifications found that body weight might be important.[2] This led to the classic description of the thin *pink puffer* and the heavier *blue bloater*.[3]

As important as these early observations were, they gave little insight into the prevalence of these abnormalities in a COPD population, the cause of the nutritional changes and the relationship to clinical outcome.

Progress in determining the prevalence of nutritional depletion in COPD was made in a series of papers published in the 1960's. In 1964, Wilson et al.[4] reported the incidence of body weight abnormalities in a group of 86 patients. From their data it is very clear that weight changes were common. Of the 86 subjects 40 were classified as underweight while only nine were overweight. Using a loss of 2.5% of original weight as the definition of weight loss, 22 patients had lost weight. Boushy et al.[5] using the criteria of a 15 pound weight loss within a three year period, found weight loss was present in approximately 25% of those patients who had three years follow-up data available. Of more importance, they found that the presence of weight loss was associated with a significantly reduced survival rate over the time studied.

Sukumalchantra and Williams[6] using a 10% weight loss to define their population found that twelve out of 44 study subjects (37%) met this criteria. Even more importantly, they found that this group had twice the mortality than those who did not lose weight (32% *vs* 16%). Vandenbergh et al.[7] corroborated these findings. Using insurance tables to determine ideal body weight and defining a low body weight as being less than ideal, 8 out of 100 patients were found to be underweight. However, they found 71% had lost 10% of their usual weight over the follow-up period. As reported by others, this weight loss was associated with a poorer prognosis.

These initial studies certainly suggest that weight changes were common and associated with a poor prognosis. In 1981, two papers[8,9] were published which gave greater insights into the incidence of nutritional changes. These two studies[8,9] evaluated other markers of malnutrition in addition to body weight and loss of weight. Several other anthropometric measures were included. These were triceps skinfolds (TSF), which reflects body fat stores and mid-arm muscle circumference (MAMC), which reflects body protein stores. Serum proteins were also analyzed as well as some immunological factors. Finally, an effort was made to look at nutritional intake.

Hunter et al.[8] evaluated 38 hospitalized patients. Sixteen of the group had lost 10% or more of their usual body weight. The Authors estimated dietary intake using a diet history; measured TSF, MAMC and the creatinine-height index; evaluated serum albumin and transferrin, and looked at the presence of delayed hypersensitivity by applying skin tests. They found, based on normal population standards, the group ingested adequate Kcal and nutrients including protein, calcium, phosphorous, iron, and several vitamins. This was true for those with a 10% weight loss also. As a matter of fact, in the COPD patients many of the ingredients were ingested in greater quantity than normal standards. Another important observation was that the group had significantly lower TSF and MAMC than standard. Albumin and transferrin were significantly reduced in comparison to expected values but were still in the low normal range. The loss of fat stores was much greater than the reduction in MAMC. Fifty percent of the group had a TSF less than 60% of standard while none of the group has MAMC reductions of that magnitude. Serum protein abnormalities were also minimal. Nine of the 32 studied were negative to all three skin tests applied (PPD, SK/SD, Candida).

Hynak et al.[9] found 42% of their study population had a severe reduction in body weight. This was defined as <80% of the Metropolitan Life tables' "ideal weight". Using the 50 percentile as standard, 29% had a TSF less than 40%, and 26% had a MAMC less than 80%. Three individuals had serum albumin less than 2.8 mg% and three had transferrin levels less than 160 mg%. A decreased MAMC seemed to be related to a reduction in the FVC percent predicted. These again were all hospitalized patients.

In another evaluation of the incidence of nutritional abnormalities, Openbrier et al.[10] divided their population into "emphysema" and "chronic bronchitis" groups.

This was done on the basis of diffusion capacity. The incidence of malnutrition was much higher in the emphysematous group. This group had a mean percent ideal body weight (IBW%) of 93 while the bronchitis group was 119%. Forty-three percent of the emphysema group had an IBW% of less than 90. While certainly the Authors' method of selection of "emphysema" is appropriate, the emphysema group also had a lower FEV_1 than the "chronic bronchitis" group. Thus one cannot separate the severity of disease from the presence of emphysema.

Several questions remain after these recent COPD population studies. Firstly, several reports are based on hospitalized patients. It is well known hospitalized medicine patients in and by themselves will have nutritional depletion.[11] Therefore it is hard to differentiate the process leading to hospitalization from the underlying pulmonary problem. Secondly, there are no real insights into the potential cause of this malnutrition. In order to answer these questions, the study to be discussed here[12] was undertaken in an outpatient population.

Purpose

The purpose of this study was:
1. to determine the prevalence of nutritional abnormalities in a stable outpatient COPD population;
2. to determine whether physiological and psychological factors were associated with these changes.

Methods

Sixty consecutive patients with COPD who were entering the *RESTOR COPD* rehabilitation program[13] were studied. All had a history of dyspnea on exertion. There were 42 men and 18 women. The mean age was 62.0 ± 1.3 years. All had FEV_1/FVC of less than 70% and remained obstructive after bronchodilators. The mean FEV_1 percent predicted for the entire population was 35.1 ± 1.9. There was no evidence of restrictive disease. Any patient with roentgenographic evidence of interstitial disease was excluded. All underwent spirometry, arterial blood gases, and diffusion capacity. They also had resting oxygen consumption (VO_2) and minute ventilation (Ve) measured. This was done after the subjects were allowed to equilibrate and found to be stable while on the mouthpiece. Nutritional assessment included measurement of body weight, TSF and MAMC. In addition whether there was a change of weight over the last year was determined. A five percent loss of weight in the last year was considered significant. Serum albumin, and pre-albumin were obtained. Body weight was expressed as percent of ideal (%IBW). Ideal was based on height, age and sex.[14] MAMC was calculated from the mid-arm circumference -0.314 x TSF (mm).[15] TSF and MAMC were then expressed as percent of standard.[16] A three day diet record was kept and reviewed with a registered dietician. Estimated Kcal requirements were made using both the Harris-Benedict equation[17] and using the resting VO_2. Psychological status was evaluated by the SCL-90R.[18] This is a series of questions which allows the determination of percentiles in a range of psychiatric categories including depression, anxiety, and a "global index". These three were used for further analysis.

For statistical analysis a correlation matrix, step-wise regressions and logistic regressions were computed on a Univac 1110.[19] Kcal ingestion was calculated on the basis of the three day diet history and expressed in two ways for further analysis. One was actual Kcal divided by predicted Kcal based on the Harris Benedict equation, or the percent basal energy expenditure (%BEE). The second was actual Kcal divided by predicted Kcal based on resting VO_2 or percent resting energy expenditure (%REE)

Results

Nutritional depletion was common, as can be seen in figure 1, 16 out of 60 subjects had lost 5% or more of their BW in the last year. Most individuals were less than 90 %IBW. TSF reductions were even more severe with approximately one-third less than or equal to 60% of standard. It is important to note that MAMC was not as severely affected. As can be seen in figure 2, there was a preservation of serum proteins. The mean serum albumin and pre-albumin were low normal. This is true for both the total population as well as the group with weight loss. Mean nutritional values for the entire group can be seen on figure 3. This indicates that overall the population's values were reasonably well preserved. However, there is a wide variation. Approximately one out of two subjects had at least one marker of nutritional depletion.

Evaluation of nutritional intake derived from the 3 day diet records revealed that the mean Kcal ingestion was much greater than normal based on the Harris-Benedict equation. (See Fig. 3, BEE% ingested) However, this relationship is much different when the Kcal requirements are calculated from the resting energy expenditure as determined by indirect calorimetry. (See Fig. 3, REE% ingested) With this calculation the % ingested closely approximates the Kcal ingested. Since these calculations are based on resting measurement, they suggest there is in-adequate Kcal ingestion for the metabolic needs of the subjects. Any increase in metabolic demand such as occurs with even a minimal increase in activity level, would place these individuals in negative energy balance.

Evaluation of the role of the various physiological parameters indicates there are no great differences, between the total group and those with weight loss (Fig. 4). This is also true for the PaO_2 and $PaCO_2$ (Fig. 5).

Table I. Nutritional and Physiologic Parameters

Nutritional	Physiological	R	P
% Ideal Body Weight	FEV$_1$ % pred.	0.25	<.05
	Diffusion % pred.	.34	<.01
	D$_L$/VA	.33	<.01
	VO$_2$/Kg	-.40	<.003

Some insight into potential reasons for nutritional depletion can be seen in Table I. There appears to be a significant relationship between markers of disease severity and nutritional changes. The % IBW is directly related to the FEV$_1$ and the diffusion capacity and inversely related to the resting $\dot{V}O_2$.

Finally Table II demonstrates the determinates of % IBW in the 28 subjects who had all the data available. Five independent variables can explain 62.61% of the variation. Most important is $\dot{V}O_2$/Kg which is inversely related. This again indi-

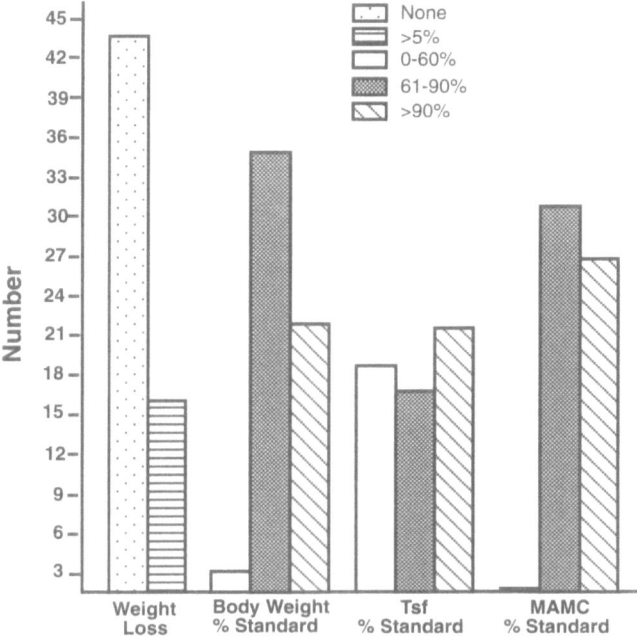

Fig. 1. The graph demonstrates an overall view of the study population. The number falling into each nutritional category is depicted.

40

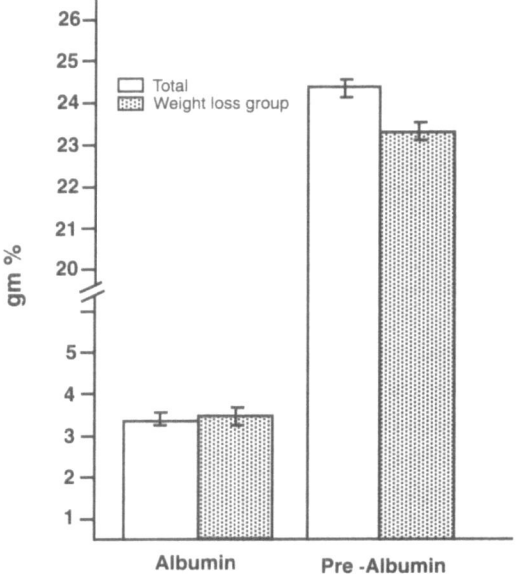

Fig. 2. Demonstration of the mean albumin (gm%) and pre-albumin (gm%) in the total population and those with weight loss.

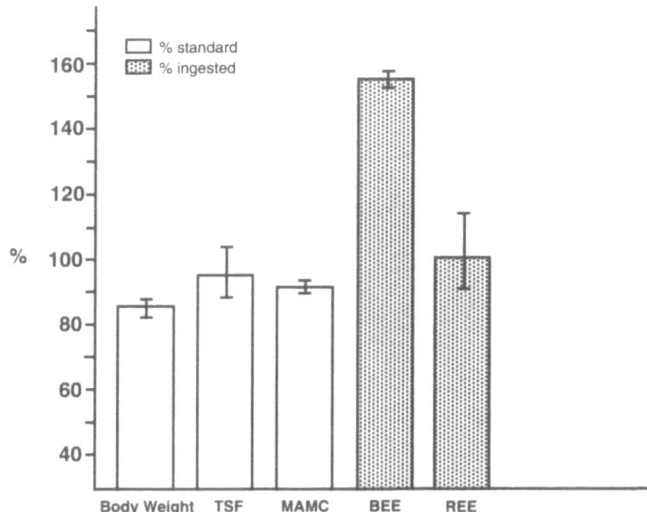

Fig. 3. Mean values of the nutritional parameters of the group as a whole. BEE% ingested represents mean Kcal actually ingested based on the three day diet history expressed as a percent basal energy expenditure calculated from the Harris-Benedict equation. %REE ingested represents mean Kcal actually ingestd based on the three day diet history expressed as a percent of Kcal needed calculated from the resting oxygen consumption.

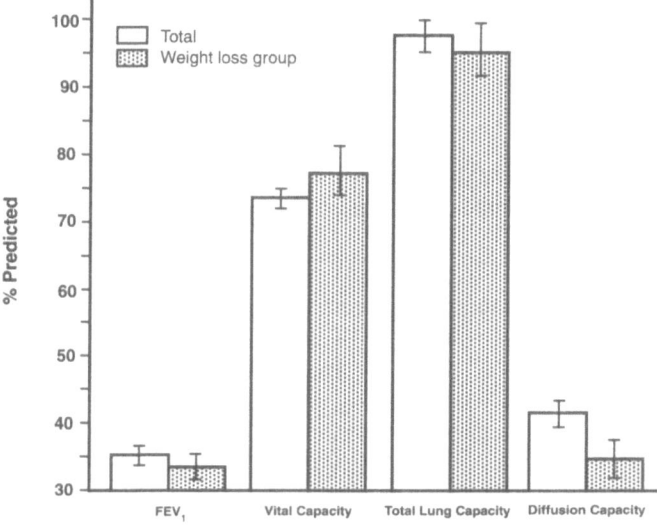

Fig. 4. Comparison of the mean values of pulmonary functions of the total group to the weight loss group.

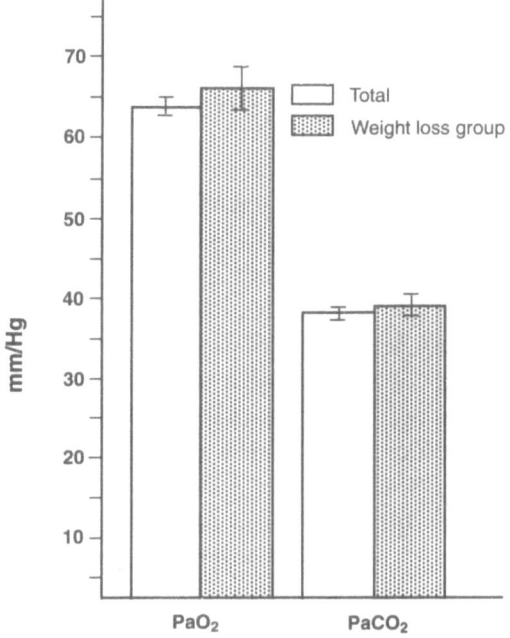

Fig. 5. Comparison of the mean PaO_2 and Pa CO_2 between the total group and those with weight loss.

cates the importance of the increasing resting energy expenditure. Also important is the efficiency of ventilation as expressed in the ventilatory equivalent ($\dot{V}e/\dot{V}O_2$).

Conclusions

These data suggest that the nutritional depletion seen in earlier studies in hospitalized COPD patients is also present in a relatively stable outpatient popula-

Table II. Determinants of nutritional parameters

Dependent	N	Independent	% Contribution
% Ideal Body Weight	28	$\dot{V}O_2/Kg$, rest (-)	22.00
		$\dot{V}e/\dot{V}O_2$ (-)	13.50
		$PaCO_2$ (-)	9.84
		Log VC% pred.	9.24
		Depression	7.79
		Total	62.61

tion. Weight loss is not as common as reduced body fat stores and low body weight. Protein status, at least as measured by serum protein and MAMC, is relatively well preserved. This suggests a marasmus type of malnutrition rather than the kwashiorkor type seen with inadequate protein intake. It appears that a relative Kcal deficiency exists.

The study presented here also indicates that an important factor in reduced body weight is increased Kcal requirement rather than decreased Kcal ingestion. This possibly is related to increased oxygen cost of breathing. Cherniack[20] reported that individuals with emphysema have marked increased costs of breathing. The data also demonstrate physiological correlates of the nutritional abnormalities. Five factors can explain almost 2/3 of the low weight. Most important is the inverse relationship between $\dot{V}O_2/Kg$ at rest and the %IBW. Also important is the inverse relationship between the ventilatory equivalent and %IBW. This indicates that these individuals with inefficient ventilation have increased Kcal needs which are only partially met by increased intake.

The data presented here still leaves unanswered questions. We still do not know whether the weight loss leads to the poor prognosis or whether it is secondary. The answer to this question has important therapeutic implications. It is not known if increased Kcal ingestion can markedly alter the prognosis even though respiratory muscle status seems to be improved. Finally, we need better insights into the role of reducing the metabolic needs by aggressive routine therapy.

References

1. Hare H. A.: Respiratory disorders of the lung. In: McCrae O. (Ed.) *Modern Medicine*. Philadelphia, Lea Febiger 1914
2. Dornhorst A.: Respiratory insufficiency. Lancet 1955; 268:1185-7
3. Filley G.F., Beckwitt H.J., Reeves J.T., Mitchell R.S.: Chronic obstructive bronchopulmonary disease II. Oxygen transport in two clinical types. Am. J. Med. 1968; 44:26-38
4. Wilson N., Wilson R.H.K., Farber S.M.: Nutrition in pulmonary emphysema. J. Am. Diet Assoc. 1964; 45:530-36
5. Boushy S.F., Adhikari P.K., Sakamoto A., Lewis B.M.: Factors affecting prognosis in emphysema. Dis. Chest 1964; 45:402-411
6. Sukumalchantra Y., Williams M.H.: Serial studies of pulmonary function in patients with chronic obstructive pulmonary disease. Am. J. Med. 1965; 39:941-945
7. Vandenbergh E., Van de Woestyne K. P., Gyselen A.: Weight changes in the terminal stages of chronic obstructive pulmonary disease. Am. Res. Respir. Dis. 1967; 95:556-65
8. Hunter A.M.B., Carey M.A., Larsh H.W.: The nutritional status of patients with chronic obstructive pulmonary disease. Am. Res. Respir. Dis. 1981; 124:376-381
9. Hynak M.T., Al-Ibrahim M.S., Russel R.M., Stanko G., Verghease C.V.J., Payton G.: Nutritional and pulmonary function assessment in patients with chronic obstructive lung disease. Nutrition Research 1981; 1:461-66
10. Openbrier D.R., Irwin M.M., Rogers R.M., Gottlieb G.P., Dauber J.H., Payton G., Van Thiel D.H., Pennock B.E.: Nutritional status and lung function in patients with emphysema and chronic bronchitis. Chest. 1983; 83: 17-22
11. Bistrian O.R., Blackburn G.L., Vitole J., Cochran D., Naylor I.: Prevalence of malnutrition in general medical patients. JAMA 1976; 235:1567-1570
12. Braun S.R., Keim N.L., Dixon R.M., Clagnaz P., Anderegg A., Shrago E.J.: The prevalence and determinants of nutritional changes in chronic obstructive pulmonary disease. Chest 1984; 86:558-563
13. Braun S.R., Driscoll S., Anderegg A., Barb J., Smith F.R., Reddan W.: A decentralized rehabilitation program for chronic obstructive airway disease in small urban and rural areas of Wisconsin: A preliminary report. Public Health Rep. 1981; 96: 315-18
14. Grande F., Keys A.: Body weight, body composition, and caloric status. In: Goodhard R.S., Shils M.E. (Eds.) *Modern Nutrition in Health and Disease*. Philadelphia, Lea Febiger, 1980; 5-6
15. Blackburn G.L., Bistrian B.R., Mairi M.S., Schlamm H.T.I., Smith M.I.: Nutritional assessment of the hospitalized patient. J.P.E.N. 1977;1:11-77
16. Bishop C.W., Bowen P.E., Ritchey S.J.: Norms for nutritional assessment of American adults by upper air anthropometry. Am. J.Clin. Nutr. 1981; 34:2530-9
17. Harris J.A., Benedict F.A.: *Biometric studies of basal metabolism in man*. Carnegie Institute of Washington, 1919. Publication No. 279
18. Derogatis L.R. The SCL-90-R. Baltimore: Clinical Psychometrics Research, 1975
19. Ryan T.A. Jr., Joiner B.L., Ryan, B.F.: *Minilab reference manual*. University Park, P.A: Minilab, Inc. January 1981
20. Cherniack R.M.: The oxygen consumption and efficiency of the respiratory muscles in health and emphysema. J. Clin. Invest. 1959; 38:494-9

5. Energy Expenditure in Chronic Lung Disease

J.W. FITTING

Division of Pneumology, Internal Medicine Department, Hospital University Centre Vaudois, Lausanne, Switzerland

Introduction

Emphysema, which is a common disease, has long been recognized to be commonly associated with weight loss in advanced cases.[39] As a consequence, most of the present knowledge on nutritional aspects in chronic lung disease relates to chronic obstructive pulmonary disease (COPD). Thus, malnutrition was frequently observed in COPD and was shown to correlate with the severity of airway obstruction.[5, 7, 12, 13, 24, 31]

Malnutrition was also shown to be consistently associated with an increased mortality rate in COPD.[4, 34, 38, 41]

The contribution of caloric intake and expenditure to weight loss in COPD was assessed by several studies and will be reviewed here. Recent studies also characterized energy expenditure in cystic fibrosis and in interstitial lung disease, thereby completing the picture of energy requirements in chronic lung disease.

Caloric Intake

Several studies indicate that a reduced caloric intake is not the primary cause of malnutrition in COPD. Vandenbergh et al.[38] evaluated the dietary intake of 50 patients with COPD, 26 with weight loss of more than 10%, and 24 without weight loss.

The caloric intake was normal in the patients with weight loss whereas it appeared too high in the patients with stable weight, suggesting that the predicted values were too low for this population. Hunter et al.[24] found that the caloric intake matched the recommended dietary allowances in 36 hospitalized COPD patients, as well as in a subgroup of 16 patients with weight loss of more than 10%. Braun et al.[5] evaluated the caloric intake of 60 outpatients with COPD and found that it exceeded the requirements based on predicted basal metabolic rate (BMR). However, caloric intake was often insufficient when compared with energy requirements based on actual oxygen consumption. Finally, the mean caloric intake was reported to be 1.4 times BMR in 6 malnourished COPD patients,[40] and 1.7 times intake of normal subjects in 13 malnourished COPD patients.[28] Thus, although nutrient intake may be reduced in some patients, these studies suggest that this is not the rule. Consequently, an increased energy expenditure was postulated as the cause of weight loss in COPD.

Energy Expenditure

In a preliminary report Heymsfield et al.[23] first described BMR to be increased to 123% of predicted in 6 malnourished COPD patients. A hypermetabolic state was then reported by three studies showing an increased resting oxygen consumption in patients with COPD.[5, 30, 40] However, these measurements were obtained using short collections of expired air with a mouthpiece, a technique which may overestimate oxygen consumption. The use of a mouthpiece and a noseclip was shown to increase tidal volume by 15-29% and minute ventilation by 14-39% .[2, 17, 19] This effect results in an increased oxygen consumption in subjects with an increased work of breathing. Thus, oxygen consumption increased by 35% in asthmatics breathing through a mouthpiece relatively to a face mask.[17] The overestimation of true resting oxygen consumption is particularly likely if measurements are taken over short periods and before a steady state has been attained.

Recently, several groups measured energy expenditure in stable COPD patients with standard methods of indirect calorimetry. Such methods include measurement of oxygen consumption and carbon dioxide production with a ventilated canopy placed over the subject's head, thereby avoiding any attachment to the face. As the subject feels comfortable with this device, the measurements can be performed over prolonged periods and are more likely to reflect the true resting metabolism. Taking into account the caloric equivalent of oxygen as determined from the respiratory quotient, the resting energy expenditure is computed in Kcal. Basal metabolic rate (BMR) refers to the energy expenditure measured under strict conditions, including absolute rest after awakening, after an overnight fast, and at thermo-neutrality. Resting energy expenditure (REE) refers to measurements performed at rest during the day without meeting the conditions of BMR, and includes diet-induced thermogenesis. Energy expenditure can be compared to that of a control group or to predicted values.[16, 22]

Goldstein et al.[20] measured energy expenditure in 10 outpatients with COPD and weight loss. During infusion of 5 percent dextrose solution, REE was increased to

116% of predicted. In contrast, REE was decreased to 90% of predicted in 5 control malnourished subjects without lung disease. The same authors then reported REE to be 113% of predicted in 10 malnourished patients with emphysema, and 116% of predicted in 9 well-nourished patients with emphysema.[21]

A similar study was performed in our laboratory where REE was measured after an overnight fast in 10 patients with COPD in stable clinical state.[14] The mean REE was increased to 117% of predicted. Because several patients were malnourished, REE was expressed per unit of fat-free mass and was compared with that of 10 control normal subjects in order to eliminate a possible effect of different body composition. Computed in this way, REE was still increased in patients with COPD, on average 25% above control subjects.

Donahoe et al.[11] reported similar results in COPD, using steady state measurements and a mouthpiece for collecting expired air. The REE was increased to 119% of predicted BMR in 10 malnourished patients whereas it was 105% of predicted BMR in 9 normally nourished patients.

Only one preliminary report on 10 patients concludes that COPD is not associated with hypermetabolism, with REE amounting to 94% of predicted in this group.[32] However, these patients were severely malnourished, with an average weight of 74% of ideal body weight. During semi-starvation a drop of approximately 40% in BMR can be expected with such a weight loss. Therefore, the REE measured in this group of patients with COPD is probably inappropriately high.

In a preliminary report on a larger number of severe COPD patients admitted to a rehabilitation program, Schols et al.[33] found an increased REE in 42 out of 84 patients. The patients showing hypermetabolism had more severe airway obstruction.

From these recent studies using standard methods of calorimetry, it can be concluded that a hypermetabolic state is frequently present in COPD. Moreover, the results from different investigators are remarkably similar.

In the absence of corresponding data for patients with interstitial lung disease, we decided to assess if this condition is associated with a hypermetabolic state as well. Twelve patients presenting a roentgenographic pattern of diffuse interstitial lung disease from different origins were studied with the same methods after an overnight fast.[15] Calculating energy expenditure both in absolute values and normalized per kg of fat-free mass, we found that REE was increased to approximately 120% of predicted, i.e. to similar values as in COPD.

Malnutrition is common in cystic fibrosis and largely results from insufficient caloric intake because of anorexia and malabsorption. Recently, two groups of investigators explored the energy expenditure of patients with cystic fibrosis in yrs. and FEV$_1$ of 61% of predicted, and found their REE increased to 119% of predicted. Buchdahl et al.[6] measured REE in 23 patients, with a mean age of 10.8 yrs. and FEV$_1$ of 83% of predicted, and found it increased to 109% of control subjects instable state. Vaisman et al.[37] studied 71 patients, with a mean age of 17.7 years and FEV_1 of 61% of predicted, and found their REE increased to 119% of predicted. Buchdahl et al.[6] measured REE in 23 patients, with a mean age of 10.8 years and FEV_1 of 83% of predicted, and found it increased to 109% of control subjects. From these two studies, it appears that REE was increased to a higher degree in patients who were older and had a more severely impaired lung function. This hypermetabolic state is likely to contribute to malnutrition in cystic fibrosis.

Thus, the assessment of metabolism by standard methods of calorimetry consist-

Table I. Resting energy expenditure in chronic lung disease.

Reference	Disease	n	REE (% pred.)
Heymsfield 1981	COPD	6	123%
Goldstein 1987	COPD	10	116%
Goldstein 1988	COPD	19	113-116%
Donahoe 1989	COPD	19	105-119%
Fitting 1989	COPD	10	117-125%
Ryan 1989	COPD	10	94%
Fitting 1990	ILD	12	117-121%
Vaisman 1987b	CF	71	119%
Buchdahl 1988	CF	23	109%

COPD: chronic obstructive pulmonary disease; ILD: interstitial lung disease; CF: cystic fibrosis.

ently revealed an increase in energy expenditure in different types of chronic lung disease, as summarized in table I.

Mechanisms of Hypermetabolism

Hypermetabolism in chronic lung disease cannot be ascribed to hyperthyroidism because thyroid hormones were measured in three studies and were found to be normal.[14, 16, 20] The influence of bronchodilator drugs cannot be always ruled out in patients with airway obstruction, as metabolic measurements were performed without interrupting therapy in some studies. Aminophylline stimulates the release of catecholamines[1] and salbutamol was shown to increase REE by 10% both in patients with cystic fibrosis and in normal subjects.[36] However, some studies were performed without bronchodilator therapy and showed hypermetabolism in COPD,[14] in interstitial lung disease,[15] and in cystic fibrosis.[37]

Currently, the most likely cause of hypermetabolism in chronic lung disease is believed to be a high work of breathing. The work of breathing is indeed increased in all these conditions, due to increased flow resistance in COPD, to decreased lung compliance in interstitial lung disease, and to both types of loads in cystic fibrosis.

The energy expended in breathing, also termed oxygen cost of breathing, can be estimated during hyperventilation. The method consists in measuring oxygen consumption and ventilation at rest and at different levels of hyperventilation. The oxygen cost of breathing is given by the slope of the relationship between oxygen consumption and ventilation, and this slope increases with increasing levels of ventilation. This method bears limitations and is fraught with problems. First, the oxygen cost of breathing is measured during hyperventilation and can only be evaluated by extrapolation for resting breathing. Second, the accuracy of the measurements is affected by the difficulty in obtaining a stable oxygen consumption, by the short duration of measurements, and by the amplified errors at higher levels of ventilation. These factors explain the wide variability of results obtained in normal subjects. Nevertheless, several studies showed that the cost of breathing during hyperpnoea is increased in patients with emphysema.[8, 9, 10, 18, 25, 26] Donahoe et al.[11] used this method by increasing ventilation only slightly. They found that the oxygen cost of breathing was increased to 21% of total oxygen consumption in malnourished COPD patients. They showed as well that this augmented cost of breathing accounted for the hypermetabolism found in these patients.

The observation that hypermetabolism is more frequent or more marked in patients with poor lung function also militates in favour of an increased work of breathing. In the study of Donahoe et al.,[11] the oxygen cost of breathing strongly correlated with the RV/TLC ratio. Schols et al.[33] reported that among 84 COPD patients those with hypermetabolism had a significantly lower FEV_1 and higher $PaCO_2$. Similarly, a significant relationship was observed between increased REE and lower FEV_1/FVC[6] or lower FEF 25-75[37] in cystic fibrosis.

Malnutrition does not affect all patients with COPD equally, some patients remaining overweight.[35] Moreover, it is commonly accepted that weight loss develops in emphysematous, or "pink puffer", patients and spares the bronchitic, or "blue bloater", patients. Openbrier et al.[29] compared patients with emphysema and with chronic bronchitis and reported that malnutrition was frequent in the former group, and absent in the latter. Unfortunately, the group with emphysema had more severe airway obstruction so that the role of emphysema per se cannot be ascertained from this study. Data obtained from a study of 779 men with COPD showed an association between malnutrition and factors representative of emphysema.[41] Thus, after adjusting for the severity of airflow obstruction by dividing the patients into three groups according to their FEV_1, body weight was related to diffusing capacity (%predicted) and inversely related to total lung capacity (% predicted).

The hypothesis that malnutrition is subserved by an increased work of breathing fits well with this observation. Indeed, patients with emphysema are characterized by marked lung hyperinflation that profoundly alters their chest wall mechanics. The inspiratory muscles are operating at a shorter than optimal length in terms of the length-tension relationship. As a consequence, they require a higher activation to generate a given tension. Moreover, the diaphragm is often considerably flattened and loses part or all of its efficacy as a pressure generator, despite high activation and muscle tension, and the inspiratory rib cage muscles are recruited to compensate for the failing diaphragm. Thus, one can easily imagine that the energy spent in breathing is higher in emphysema than in chronic obstructive bronchitis, which is less frequently associated with hyperinflation. However, we found no relationship between REE and various indices reflecting chest wall distortion in 10 patients with stable COPD.[14] Thus, the actual role of lung hyperinflation in generating hypermetabolism has yet to be determined.

Finally, Goldstein et al.[21] characterized the nutrient oxidation in patients with COPD and hypermetabolism. These authors discovered that the increased REE of COPD patients was due to higher carbohydrate oxidation, whereas hypermetabolism in septic and injured patients is characterized by predominant fat oxidation.[2] This observation lends additional support to the hypothesis that the extra energy is expended by the respiratory muscles.

In summary, several recent studies demonstrated an increased resting energy expenditure in chronic lung disease of different origins. This extra energy is most likely spent by the respiratory muscles facing an increased load.

The hypermetabolism appears to increase with the severity of lung disease and contributes to the weight loss frequently observed in these disorders.

References

1. Atuk N.O., Blaydes M.C., Westervelt F.B., Wood J.E.: Effect of aminophylline on urinary excretion of epinephrine and norepinephrine in man. Circulation 1967; 35: 745-753
2. Askanazi J., Carpentier Y.A., Elwyn D.H., Nordenström J., Jeevanandam M., Rosembaum S.H., Gump F.E., Kinney J.M.: Influence of total parenteral nutrition on fuel utilization in injury and sepsis. Ann. Surg. 1980; 191: 40-46
3. Askanazi J., Silverberg P.A., Foster R.J., Hyman A.I., Milic-Emili J., Kinney J.M.: Effects of respiratory apparatus on breathing pattern. J. Appl. Physiol. 1980; 48: 577-580
4. Boushy S.F., Adhikari P.K., Sakamoto A., Lewis B.M.: Factors affecting prognosis in emphysema. Dis. Chest 1964; 45:402-411
5. Braun S.R., Keim N.L., Dixon R.M., Clagnaz P., Anderegg A., Shrago E.S.: The prevalence and determinants of nutritional changes in chronic obstructive pulmonary disease. Chest 1984; 86: 558-563
6. Buchdahl R.M., Cox M., Fulleylove C., Marchant J.L., Tomkins A.M., Brueton M.J., Warner J.O.: Increased resting energy expenditure in cystic fibrosis. J. Appl. Physiol. 1988; 64: 1810-1816
7. Burrows B., Niden A.H., Barclay W.R., Kasik J.E.: Chronic obstructive lung disease. II. Relationship of clinical and physiologic findings to the severity of airways obstruction. Am. Rev.

Respir. Dis. 1964; 91: 665-678

8. Campbell E.J.M., Westlake E.D., Cherniack R.M.: Simple methods of estimating oxygen consumption and efficiency of the muscles of breathing. J. Appl. Physiol. 1957; 11:303-308

9. Cherniack R.M.: The oxygen consumption and efficiency of the respiratory muscles in health and emphysema. J. Clin. Invest. 1959; 38: 494-499

10. Cournand A., Richards D.W., Bader R.A., Bader M.E., Fishman A.P.: The oxygen cost of breathing. Trans. Assoc. Am. Phys. 1954; 67:162-173

11. Donahoe M., Rogers R.M., Wilson D.O., Pennock B.E.: Oxygen consumption of the respiratory muscles in normal and in malnourished patients with chronic obstructive pulmonary disease. Am.Rev. Respir. Dis. 1989; 140: 385-391

12. Driver A.G., McAlevy M.T., Smith J.L.: Nutritional assessment of patients with chronic obstructive pulmonary disease and acute respiratory failure. Chest 1982; 82: 568-571

13. Fiaccadori E., Del Canale S.,Coffrini E., Vitali P., Antonucci C., Cacciani G., Mazzola I., Guariglia A.: Hypercapnic-hypoxemic chronic obstructive pulmonary disease (COPD): influence of severity of COPD on nutritional status. Am. J. Clin. Nutr. 1988; 48: 680-685

14. Fitting J.W., Frascarolo P., Jéquier E, Leuenberger P.: Energy expenditure and rib cage-abdominal motion in chronic obstructive pulmonary disease. Eur. Respir. J.: 1989; 2: 840-845

15. Fitting J.W., Frascarolo P., Jéquier E., Leuenberger P.: Resting energy expenditure in interstitial lung disease. Am. Rev. Respir. Dis. 1990; 142: 631-635

16. Fleisch A.: Le metabolisme basal standard et sa détermination au moyen du "Metabocalculator". Helv. Med. Acta. 1951; 18: 23-44

17. Freedman A.R., Lavietes M.H.: Energy requirements of the respiratory musculature in asthma. Am. J. Med. 1986; 80:215-222

18. Fritts H.W., Filler J., Fishman A.P., Cournand A.: The efficiency of ventilation during voluntary hyperpnea: studies in normal subjects and in dyspneic patients with either chronic pulmonary emphysema or obesity. J. Clin. Invest. 1959; 38:1339-1348

19. Gilbert R., Auchincloss J.H., Brodsky J., Boden W.: Changes in tidal volume, frequency, and ventilation induced by their measurement. J. Appl. Physiol. 1972; 33: 252-254

20. Goldstein S.A., Askanazi J., Weissman C., Thomashow B., Kinney J.M.: Energy expenditure in patients with chronic obstructive pulmonary disease. Chest. 1987; 91: 222-224

21. Goldstein S.A., Thomashow B.M., Kvetan V., Askanazi J., Kinney J.M., Elwyn D.H.: Nitrogen and energy relationships in malnourished patients with emphysema. Am. Rev. Respir. Dis. 1988; 138: 636-644

22. Harris J.A., Benedict F.G.: A biometric study of basal metabolism in man. Carnegie Institute of Washington ,1919

23. Heymsfield S.B., Head A., Grossman G., Staton G.W.: Mechanisms of cachexia in chronic obstructive pulmonary disease (abstract). J. Parenteral. Enteral. Nutr. 1981; 5: 562

24. Hunter A.M.B., Carey M.A., Larsh H.W.: The nutritional status of patients with chronic obstructive pulmonary disease. Am. Rev. Respir. Dis. 1981; 124: 376-381

25. McGregor M., Becklake M.R.: The relationship of oxygen cost of breathing to respiratory mechanical work and respiratory force. J. Clin. Invest. 1961; 40: 971-980

26. McKerrow C.B., Otis A.B.: Oxygen cost of hyperventilation. J .Appl. Physiol. 1956; 9: 375-379

27. Mitchell R.S., Filley G.F.: Chronic obstructive bronchopulmonary disease. I. Clinical features. Am. Rev. Respir. Dis. 1964; 89: 360-371

28. Nørregaard O., Tottrup A., Saaek A., Bisballe S., Hessov I.: Nutritional intake in malnourished patients with chronic obstructive pulmonary disease (abstract). Eur. Respir. J. 1989; 2: 364s

29. Openbrier D.R., Irwin M.M., Rogers R.M., Gottlieb G.P., Dauber J.H., Van Thiel D.H., Pennock B.E.: Nutritional status and lung function in patients with emphysema and chronic bronchitis.

Chest 1983; 83:17-22

30. Openbrier D.R., Irwin M.M., Dauber J.H., Owens G., Rogers R.M.: Factors affecting nutritional status and the impact of nutritional support in patients with emphysema. Chest 1984; 85: 67s-69s

31. Renzetti A.D., McClement J.H., Litt B.D.: The veterans administration cooperative study of pulmonary function. III. Mortality in relation to respiratory function in chronic obstructive pulmonary disease. Am. J. Med. 1966; 41:115-129

32. Ryan C.F., Buckley P., Road J., Ross S., Whittaker J.S.: Energy balance in stable malnourished COPD patients (abstract). Am. Rev. Respir. Dis. 1989; 139 (4): A333

33. Schols A.M.W.J., Mostert R., Soeters P.B., Greve L.H., Wouters E.F.M.: Resting energy expenditure in COPD (abstract). Eur. Respir. J. 1989; 2:359s

34. Sukumalchantra Y., Williams M.H.: Serial studies of pulmonary function in patients with chronic obstructive pulmonary disease. Am. J. Med. 1965; 39:941-945

35. Tirlapur V.G., Mir M.A.: Effect of low caloric intake on abnormal pulmonary physiology in patients with chronic hypercapneic respiratory failure. Am. J. Med. 1984; 77:987-994

36. Vaisman N., Levy L.D., Pencharz P.B., Tan Y.K., Soldin S.J., Canny G.J., Hahn E.: Effect of salbutamol on resting energy expenditure in patients with cystic fibrosis. J. Pediatr. 1987a; 111: 137-139

37. Vaisman N., Pencharz P.B., Corey M., Canny G.J., Hahn E.: Energy expenditure of patients with cystic fibrosis. J. Pediatr. 1987b; 111:496-500

38. Vandenbergh E., Van de Woestijne K.P., Gyselen A.: Weight changes in the terminal stages of chronic obstructive pulmonary disease. Am. Rev. Respir. Dis. 1967; 95:556-565

39. Wilson D.O., Rogers R.M., Hoffman R.M.: Nutrition and chronic lung disease. Am. Rev. Respir. Dis. 1985; 132: 1347-1365

40. Wilson D.O., Rogers R.M., Sanders M.H., Pennock B.E., Reilly J.J.: Nutritional intervention in malnourished patients with emphysema. Am. Rev. Respir. Dis. 1986; 134:672-677

41. Wilson D.O., Rogers R.M., Wright E.C., Anthonisen N.R.: Body weight in chronic obstructive pulmonary disease. The National Institutes of Health intermittent positive-pressure breathing trial. Am. Rev. Respir. Dis. 1989; 139:1435-1438

6. Polyunsaturated Fatty Acids as Nutritional Modulators of the Pulmonary Response to Sepsis

M. J. MURRAY

Critical Care Service, Department of Anesthesiology, Nutritional Support Service, Department of Internal Medicine, Mayo Clinic and Mayo Foundation Rochester, Minnesota, USA

Introduction

The salutary effects of nutritional therapy on respiratory function are receiving increased attention. However, caution must be exercised since the nutrients we supply, especially with respect to fatty acids, not only can be oxidized to meet energy requirements, but have important pharmacologic effects as well.

Humans have the capacity to manufacture saturated fatty acids in the liver using malonyl CoA as a carbon source. When these fatty acids are dehydrogenated, hepatic enzymes are incapable of desaturating the carbon skeleton any closer to the terminal end of the molecule than at the omega (terminal) 9 position. For example, stearic acid (18.0) can be desaturated to oleic acid ($18:1\omega9$), but not to linoleic acid ($18:2\omega6$) or alpha-linolenic acid ($18:3\omega3$). The latter two compounds are essential fatty acids and therefore, must be ingested. They serve as precursors for two important compounds, arachidonic ($20:4\omega6$) and eicosapentaenoic ($20:5\omega3$) acid.

Arachidonic acid is a 20-carbon chain fatty acid with four double bonds. Counting from the terminal end of the molecule, its fourth double bond is on the sixth carbon atom, leading to the following nomenclature for arachidonic acid: $20:4\omega6$.

Arachidonic acid is stored in the phospholipids of cell membranes, is released in response to a variety of stimuli, and is metabolized to a group of dienoic compounds which include the "two-series" of prostaglandins and thromboxanes (e.g., Prostaglandin I_2, PGI_2, and Thromboxane A_2, TxA_2), important compounds which mediate regional blood flow and the inflammatory response in a variety of organs.

There is cause for concern because the currently available intravenous fat emulsions (IVFE) contain high amounts (50-60%) of linoleic acid ($18:2\omega6$), the precursor for arachidonic acid. Originally, IVFE were given once or twice a week to avoid the sequelae of essential fatty acid deficiency. More recently, many authors advocate the administration of IVFE on a daily basis as a caloric substrate. There is concern that the large amount of infused linoleic acid may be driving the arachidonic acid cascade with possible deleterious effects on organ function.

Pulmonary Effects of IVFE

Reports of adverse sequelae of IVFE first appeared in the 1960s when "Lipomul" was reported to induce a disease state similar to the traumatic fat embolism syndrome. However, even with reformulation of products, reports continued to appear regarding the adverse effects of IVFE on pulmonary function.[1]

Other studies have reported that IVFE affect lung function by decreasing the carbon monoxide diffusing capacity (DLCO) with the development of hypoxia.[2] The decrease in DLCO was originally thought secondary to the mechanical effects of the lipid chylomicrons and triglycerides on the pulmonary vascular bed. However, these studies have emphasized the transient nature of the changes and have concluded that they are not clinically relevant.

More recently, there has been concern that perhaps eicosanoid compounds, augmented by the infused IVFE, may be responsible for the hypoxia that has been seen.[3] It has long been known that prostaglandins and thromboxanes play an important role in the lungs with important effects on both the pulmonary vasculature and airways.[4,5] They are equally important in the development of pulmonary pathophysiology in a variety of disease states, such as sepsis and acute lung injury. Patients with these syndromes are the same ones who are often receiving IVFE. Are there any similarities?

Halushka and Brigham have long emphasized the role of prostanoid compounds in the development of sepsis and acute lung injury, respectively.[6,7] Sheep injected with endotoxin have an increase in pulmonary vascular resistance with the development of hypoxia and interstitial pulmonary oedema, thought secondary to the increased intravascular hydrostatic pressure.[8] An identical syndrome can be initiated by infusing the sheep with IVFE.[9] The importance of thromboxane as the causative agent has been underscored by the findings of elevated plasma thromboxane levels following the administration of endotoxin. In addition, pulmonary response to endotoxin can be attenuated by the administration of indomethacin, a cyclooxygenase inhibitor which blocks the formation of thromboxane.[10]

In our laboratory, we undertook a series of experiments to explore the relationship between IVFE, prostanoid formation, and effects on the respiratory system.[11] In one series using an unanaesthetized porcine model, we infused pigs with Intralipid at a rate of 0.25 gm. kg^{-1}. hr^{-1}.

This corresponds (given the pig's metabolic rate and body size difference) to a daily dose of 3 gm. kg⁻¹ for humans, the maximum dose recommended. There was an increase in pulmonary vascular resistance and pulmonary arterial pressure, and hypoxia developed transiently (Fig.1). As had been reported previously, the hypoxia was transient, resolving within one hour.

However, since there is particular concern regarding the use of IVFE in stressed patients, a dose of endotoxin was chosen (10 ng. kg⁻¹ I.V. bolus followed by 8 ng

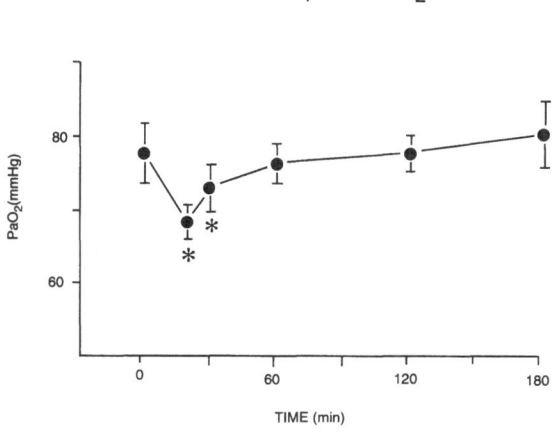

Fig. 1. Effects of intravenous Intralipid (o.24 gm . kg⁻¹ . hr ⁻¹) on PaO₂ in a group of 6 pigs.
 * p < 0.05 by Student's t-test, compared to baseline.

. kg⁻¹ . hr ⁻¹) which produced no effects on the PaO₂. Pigs were then assigned to receive, in a crossover design, either endotoxin or endotoxin and Intralipid for 3 hours, followed 24 hours later with the animal assigned to receive the opposite infusion. As was hypothesized, the combination of endotoxin and Intralipid produced a sustained decrease in PaO₂ which persisted for the 3 hours of the experiment (Fig. 2). While the decrease in PaO₂ was small (approximately 10 mmHg), such differences over time may, indeed, have a significant impact on outcome. Furthermore, it should be emphasized that the dose of endotoxin used in these experiments was of a magnitude to produce no apparent cardiopulmonary compromise. What would be the effects of IVFE in situations where the stress was much greater?

Another important result of these studies was the observation that thromboxane B₂ (TxB₂), the stable metabolite of thromboxane A₂ (TxA₂), increased significantly in the group of pigs receiving endotoxin plus Intralipid (Fig. 3) compared to the

Effects of Intralipid and Endotoxin on PaO$_2$

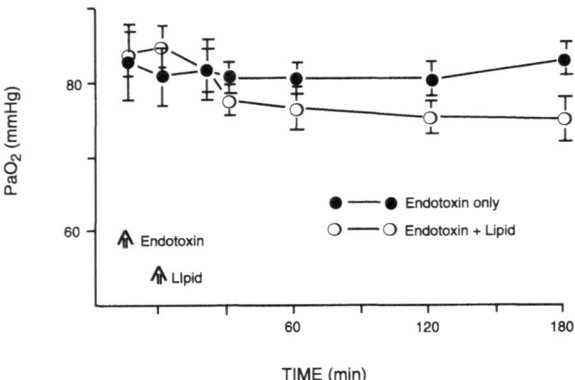

TIME (min)

Fig. 2 Effects of endotoxin on PaO$_2$ in a group of pigs (n = 6) compared to another group of pigs (n = 6) receiving the same dose of endotoxin with the addition of Intralipid. Endotoxin (E. coli) was given intravenously at time t = -15 minutes as a 10 ng . kg^{-1} bolus followed by an infusion of 8 ng . kg^{-1} . hr^{-1}. In the group of pigs receiving Intralipid, it was given intravenously beginning at time t = 0 at a rate of 0.25 gm . kg^{-1} . day^{-1}.

Comparing the area under the curve (AUC) between the two groups, the group receiving endotoxin plus Intralipid had significantly lower PaO$_2$ values (p < 0.05).

Effects of Intralipid and Endotoxin on TxB$_2$ Levels

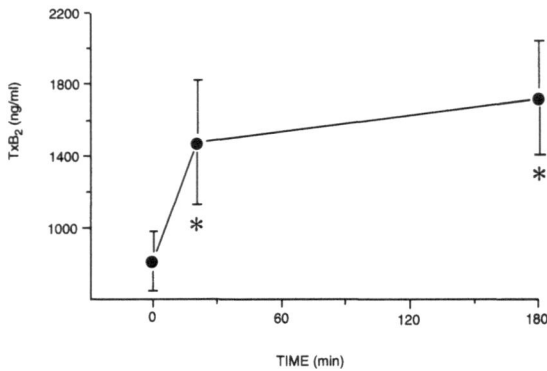

TIME (min)

Fig. 3 Thromboxane B$_2$, the stable metabolite of thromboxane A$_2$, was measured at baseline and at 20 minutes and 3 hours following the administration of Intralipid and endotoxin. p < 0.05 compared to baseline.

group receiving endotoxin alone and was statistically associated with the fall in PaO$_2$. These results indicate both the possible mechanism as well as the pulmonary side effects of IVFE.

It can only be speculated as to why an elevation of prostanoids should lead to the development of hypoxia. Some studies have suggested that increased prostaglandin levels may actually be beneficial in acute lung injury. Shoemaker and Appel have demonstrated that PGE_1, a vasodilating prostaglandin may be beneficial in the adult respiratory disease syndrome.[12] PGI_2, on the other hand, also a pulmonary vasodilating prostaglandin, has been associated with the development of hypoxia in animal models.[13]

The answer to this paradox may become apparent if one examines the role that the prostaglandins play in the control of the regional lung vascular bed. TxA_2 is a vasoconstrictor, PGI_2 is a vasodilator, and the balance between the two may control blood flow to different regions of the lung. In well-ventilated areas, an excess of TxA_2 may lead to vasoconstriction with an increase in the pulmonary dead space (VD/VT). An increase in PGI_2 in an area of poor ventilation could lead to an increased shunt with a decrease in PaO_2. An elevation of these eicosanoids, therefore depending on the vascular bed, could easily lead to an increase in the pulmonary shunt fraction or pulmonary dead space, with an associated abnormality in gas exchange. As the final part of these experiments, we studied another group of pigs infusing the same dose of endotoxin, but infusing the Intralipid at only 0.08 $kg^{-1} . hr^{-1}$ (corresponding to a dose of 1 gm . $kg^{-1} . day^{-1}$) in humans. No effects on PaO_2 nor on the pulmonary vasculature were observed.

Our conclusion is that patients receiving IVFE should not receive 3 gm . kg^{-1} . day^{-1}. A total dose of 1 gm . $kg^{-1} . day^{-1}$ is acceptable if it is infused over 18 to 24 hours. Shorter infusion periods increase the risk of developing prostanoid mediated alterations in pulmonary arterial pressure and hypoxia. Studies to prove this in humans should be considered, but are difficult to perform, primarily because it is difficult to control the underlying disease process. In addition though, given the magnitude of the change in PaO_2 and the intrinsic variability of PaO_2 measurements (approximately 4%) in order to demonstrate a significant difference between infusion rates at the 90% confidence level, a very large number of patients (greater than 50 in each group) would have to be studied. Given the currently available literature, however, there is little reason to give IVFE at a faster rate or greater dose than what is recommended (0.04 gm . $kg^{-1} . hr^{-1}$ or 1 gm . $kg^{-1} . day^{-1}$).

Pulmonary Effects of ω3 IVFE

If IVFE can be safely given at an appropriate rate and dose, is there any reason to consider alternate fat sources for parenteral nutrition? The answer is yes for a number of reasons. The long chain fatty acids (18 to 20 carbon chain fatty acids) contained in IVFE require L-carnitine for transport across mitochondrial membranes for oxidation. Since there is a deficiency of cellular carnitine in a number of disease states, fatty acids (even if given in an appropriate dose) may still lead to hepatic steatosis and hypertriglyceridemia. A number of investigators are involved in studies of medium-chain triglycerides as an alternate lipid fuel source for parenteral nutrition.[14]

Longer chain fatty acids still must be infused though to avoid the sequelae of essential fatty acid deficiency. While a dose of ω6 fatty acids could be easily chosen to meet these requirements, the dose of ω3 fatty acids which would be necessary is less well defined. Furthermore, there is a growing body of evidence that ω3 fatty acids may actually be beneficial in certain stress states, by modifying the eicosanoid precursor pool.

The salutary effects of ω3 fatty acids were first appreciated when investigators sought to explain the epidemiologic findings that natives of Greenland had a low incidence of atherosclerotic disease.[15] Several studies have revealed that the ω3 fatty acids contained in cold water fish are responsible for these effects, by altering triglyceride levels[16] or by altering the types[17] and/or levels[18] of eicosanoids produced. Eicosapentaenoic acid (EPA) is a 20-carbon chain fatty acid with five double bonds, with the fifth bond in the ω3 position (20:5ω3). It competes with arachidonic acid for storage in membrane phospholipids and can be released by phospholipase A_2 in response to an appropriate stimulus (stress). It serves as a substrate for cyclooxygenase and lipoxygenase, leading to trienoic prostaglandins (PGI_3, TxA_3), and leukotrienes of the five series. The ingestion of ω3 fatty acids, and in particular EPA, has been associated with an improvement in a number of disease states such as hypertension,[19] psoriasis,[20] and sepsis.[21]

A series of experiments were performed to assess the effects of ω3 fatty acids in acute lung injury and sepsis.[22] Pigs were fed for 8 days diets that were equal in all respects except for their lipid composition. Diets were supplemented with either menhaden oil (ω3 fatty acid diet) or corn oil (ω6 fatty acid diet).

The ω3 fatty acid diet contained 11% alpha linolenic acid and 1% EPA, both ω3 fatty acids and approximately 11% ω6 fatty acid. The ω6 fatty acid contained 45% linoleic acid and 1% arachidonic acid and had less than 1% ω3 fatty acid. On the ninth day, pigs were placed unanaesthetized in a standing harness and injected either with endotoxin or live bacteria. Endotoxin or live bacteria produced an immediate rise in pulmonary artery pressure (greater than 100% increase) with a drop in PaO_2 (greater than 20 mmHg). The ω3 fatty acids significantly attenuated these changes (Fig. 4). Furthermore, following the infectious insult, there was an order of magnitude increase in TxB_2 levels. These increases were statistically associated with the fall in PaO_2 (the greater the TxB_2 levels, the greater the fall in PaO_2), and again the group consuming the ω3 fatty acid enriched diet had a significant attenuation of the rise in TxB_2. In the small group of animals studied (n=8 in each group), there was no difference in outcome for any of the animals. In support of our findings, others have likewise demonstrated a beneficial effect from ω3 fatty acids in sepsis.[21]

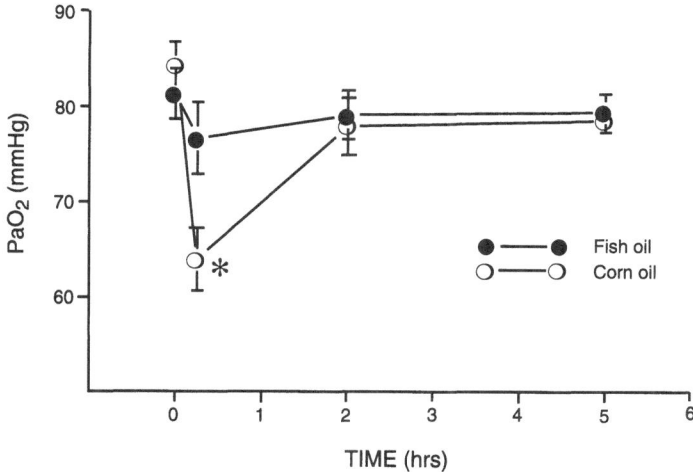

Fig. 4. The effects of endotoxin on PaO_2 over time. Two groups of pigs were fed for 8 days a fish or corn oil diet and then infused with a bolus of intravenous endotoxin (0.3 mg . kg-1) at time t = 0. $\mathbf{x} \pm$ SEM. * p < 0.05.

A cautionary note must be raised, however. In a group of pigs (n = 8) given a 10% ω3 fatty acid emulsion intravenously (0.25 gm . kg^{-1} . min^{-1}), there was a significant fall in PaO_2 over a 4 hour infusion period.

Perhaps the high degree of unsaturation that the ω3 fatty acid emulsion had might explain the observed results. Of note though is that this group, as had been found in the orally fed animals, had an improved response to endotoxin compared to a group of animals receiving intralipid, an IVFE with high concentrations of ω6 fatty acids. Further studies need to be done to delineate the role of these fatty acids in the pulmonary response to injury. These studies could be extended to explore the relationship of these dietary lipids to reactive airway diseases based on clinical studies which have demonstrated a possible beneficial role for these fatty acids in asthmatics.[23]

Dietary Influence on Pulmonary Surfactant

A final point should be discussed regarding the effects of lipids on the lung, and that is that they potentially may alter pulmonary surfactant composition.

It has long been known that the diet can affect surfactant. There is evidence that starvation can alter the fatty acid composition and increase the surface tension of pulmonary surfactant.[24] It is more difficult to demonstrate that a surplus of fatty acids impact on the composition of surfactant. This is because the majority of the phospholipids present in alveolae is endogenously produced.[25] However, few studies have assessed the effects of dietary lipids in stressed states. In the previously described experiments which examined the effects of altering the fatty acid composition of dietary lipids for 8 days, we found that during stress exogenous dietary fatty acids were incorporated into the pulmonary surfactant lavaged from the lungs at the end of the experiment (Table I). The differences were statistically significant, but had no effect on the pulmonary compliance (Fig. 5).

Conclusion

The dietary lipids we feed our pulmonary patients not only serve as an energy source with the potential to decrease carbon dioxide production, but may serve as pharmacologic mediators of the pulmonary vasculature and airways. If IVFE are given to patients, they must be given at an acceptable rate and dose. Furthermore, preliminary evidence suggests that altering the ratio of ω3 to ω6 fatty acids in lipid emulsions may actually protect the lungs against the sequelae of nosocomial infections and injury.

Table I. The fatty acid composition of the surfactant phospholipids, and of the phosphatidylcholine and phosphatidylglycerol component of pulmonary surfactant. Two groups of pigs had been fed ω3 vs. ω6 supplemented diets for 8 days prior to the development of septic shock. After 6 hours the lungs were lavaged and surfactant fatty acid composition measured by gas liquid chromatography.

	SAT FA[a]	MONOCAPS[b]	PUFA[c]	18.2ω6	20.4ω6	20.5ω3
Total Phospholipids						
ω3 Group	70.1±2.5	22.4±1.1	7.5±2.4**	3.2±0.9**	0.7±0.4	0.9±0.8
ω6 Group	65.9±5.1	21.5±5.0	12.6±2.2	9.0±3.2	1.2±0.7	0.2±0.2
Phosphatidyl choline						
ω3 Group	75.0±1.2	19.5±0.4	5.6±1.5*	2.34±0.6**	0.3±0.2*	1.0±0.5**
ω6 Group	72.4±3.3	17.1±2.6	10.5±2.6	8.6±2.3	0.8±0.4	0.1±0.02
Phosphatidyl glycerol						
ω3 Group	48.0±2.4	34.1±4.1	17.9±3.4	5.9±0.7*	2.3±2.1	2.1±2.4
ω6 Group	47.6±17.2	30.8±9.5	21.5±7.9	13.2±4.3	3.8±2.9	0.28±0.13

Data are plotted as percent of total fatty acids as the mean ± SD. * $p < 0.05$, ** $p < 0.01$. [a] Saturated fatty acids; [b] monounsaturated fatty acids; [c] polyunsaturated fatty acids.

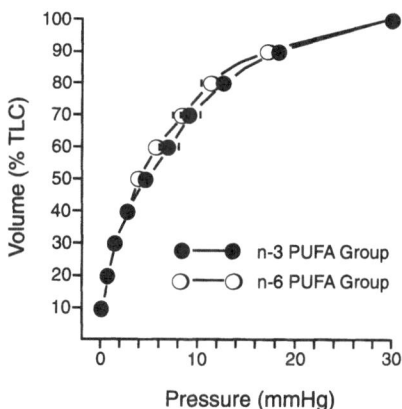

Fig. 5. Pressure-volume curves for two groups of pigs (n = 8) which received for 8 days either an ω3 or an ω6 fatty acid supplemented diet and were then subjected to bacteremic shock for 6 hours. There were no differences between the two groups. The curves were measured with air, inflating to total lung capacity (TLC) and deflating at 10% of TLC intervals. Data plotted as mean ± SEM.

References

1. Barson A.J., Chiswick M.L., Doig C.M.: Fat metabolism in infancy after intravenous fat infusions. Arch. Dis. Child 1978; 53:218-223
2. Sundstrom, Zauner C.W., Arborelius M. Jr.: Decrease in pulmonary diffusing capacity during lipid infusion in healthy men. J. Appl. Physiol. 1973; 34:816-820
3. Hageman J.R., McCulloch K., Gora P., Olsen E.K., Pachman L., Hunt C.E.: Intralipid alterations in pulmonary prostaglandin metabolism and gas exchange. Crit. Care Med. 1983; 11:794-798
4. Voelkel N.F., Chang S.W., McDonnel T.J., et al.: Role of membrane lipids in the control of normal vascular tone. Am. Rev. Respir. Dis. 1987; 136:214-217
5. Spannhake E.W., Hyman A.L., Kadowitz P.J.: Dependence of the airway and pulmonary vascular effects of arachidonic acid upon route and rate of administration. J. Pharmacol. Exp. Ther. 1980; 212:584-590
6. Halushka P.V., Wise W.C., Cook J.A.: Protective effects of aspirin in endotoxic shock. J. Pharmacol. Exp. Ther. 1981; 218:464-469
7. Brigham K.: Metabolites of arachidonic acid in experimental lung vascular injury. Fed. Proc. 1985; 44: 43-45
8. Brigham K., Bowers R., Haynes J.: Increased sheep lung vascular permeability caused by E. Coli endotoxin. Circ. Res. 1979; 45:292-297
9. McKeen C.R., Brigham K.L., Bowers R.E., Harris T.R.: Pulmonary vascular effects of fat emulsion infusion in unanesthetized sheep. Prevention by indomethacin. J. Clin. Invest. 1978; 61:1291-1297
10. Hanly P.J., Roberts D., Dobson K., Light R.B.: Effect of indomethacin on arterial oxygenation in critically ill patients with severe bacterial pneumonia. Lancet 1987;1:351-354
11. Beers T.R., Murray M.J., Miles J.M.: Intralipid induces prostaglandin synthesis and hypoxemia in the endotoxemic pig. JPEN 1990; 14: 135 (manuscript in press) Am. J. Clin. Nutr.
12. Shoemaker W.C., Applel P.L.: Effects of prostaglandin E$_1$ in adult respiratory distress syndrome.

Surgery 1986; 99:275-283

13. Sprague R.S., Stephenson A.H., Lonigro A.J.: Prostaglandin I_2 supports blood flow to hypoxic alveoli in anesthetized dogs. J. Appl. Physiol. 1984; 56:1246-1251

14. Beaufrère B., Tessari P., Cattalini M., Miles J., Haymond M.W.: Apparent decreased oxidation and turnover of leucine during infusion of medium chain triglycerides. Am. J. Physiol. 1985; 249: E175-E182

15. Dyerberg J., Bang H.O., Stoffersen E., et al.: Eicosapentaenoic acid and prevention of thrombosis and atherosclerosis? Lancet 1979;1:433-435

16. Von Schacky C., Fischer S., Weber P.C.: Long-term effects of dietary marine ω-3 fatty acids upon plasma and cellular lipids, platelet function, and eicosanoid formation in humans. J. Clin. Invest. 1985: 76:1626-1631

17. Needleman P., Raz A., Minkes M.S., Ferrendelli J.A., Sprecher H.: Triene prostaglandins. prostacyclin and thromboxane biosynthesis and unique biological properties. Proc. Natl. Acad. Sci. USA 1979: 76:944-948

18. Fischer S., Weber P.C.: The prostacyclin/thromboxane balance is favourably shifted in Greenland Eskimos. Prostaglandins 1986; 32:235-241

19. Singer V.P., Wirth M., Mest H-J., Taube C., Richter-Heinrich E., Codicke W., Hartrodt W., Naumann E., Voigt S.: Anderungen von Blutdruck und Serumlipiden unter Fischdiäten bei Patienten mit milder essentieller Hypertonie. Z. Gesamte. Inn. Med. 1986; 41:38-44

20. Greaves M.W., Sondergaard J., McDonald-Gibson W.: Recovery of prostaglandins in human cutaneous inflammation. Brit. Med. J 1971; 2:258-60

21. Mascioli E.A., Iwasa Y., Trimbo S., et al.: Endotoxin challenge after menhaden oil diet. Effects on survival of guinea pigs. Am. J. Clin. Nutr. 1989; 49:277-282

22. Murray M.J., Svingen B.A., Yaksh T.L.: The effects of a fish oil diet on pigs' cardiopulmonary response to bacteremia. (In press) J. Parenter. Enteral. Nutr., 1991

23. Arm J.P., Horton C.E., Mencia-Huerta J-M, House F., Eiser N.M., Clark T.J.H., Spur B.W., Lee T.H.: Effect of dietary supplementation with fish oil lipids on mild asthma. Thorax 1988; 43:84-92

24. Garbagni R., Coppo F., Grassini G., Cardellino G.: Effects of lipid loading and fasting on pulmonary surfactant. Respiration 1968; 25:458-464

25. Patterson C.E., Davis K.S., Rhoades R.A.: Regulation of fetal lung disaturated phosphatidylcholine synthesis by de novo palmitate supply. Biochimica et Biophysica Acta 1988: 958:60-69

Effects of Malnutrition on Ventilatory System

7. Nutritional Status and Respiratory Mechanics

N.M.T. Braun

Department of Medicine, Pulmonary Division, St. Luke's-Roosevelt Hospital Center
Clinical Medicine, College of Physicians and Surgeons, Columbia University, New York,
USA

Malnutrition is a disorder of nutrition due either to faulty or inadequate dietary intake, which results in an abnormality in body mass or composition.

Criteria have been established to define this state. Obesity has been defined as a body weight or mass index (i.e. kilograms divided by the height in meters squared) greater than 20% of ideal (males > 27.8 kg/m², for females 27.3 kg/m²).[14,21] Weight loss is defined by body weight reduction equal to or less than 90% of ideal body weight. Both conditions give rise to changes of chest wall and lung mechanics which alter ventilatory function, and increase morbidity and mortality. While equally important, I will only briefly review the consequences of obesity as it will be addressed elsewhere.

Two juxtaposed postero-anterior chest roentgenograms are used to illustrate the striking effect of weight on the thorax. (Fig. 1). The upper panel is from a 24 year old woman with anorexia nervosa with 72% IBW. The lower panel is from an obese woman who is 193% of IBW. Both films are at maximal inspiration, but the anorexic subject has more than twice the air volume as seen by the very low diaphragms and increased rib distances. The compliance characteristics of these two thoraces must be different.

Consequences of Excess Weight

Obesity is associated with increased risk for ischemic heart disease, hypertension and consequent renal failure, and stroke. (Table I) Metabolic and endocrine disorders including diabetes mellitus, gout and hypogonadism are more common. There is also an increased rate of gall bladder disease, hiatus hernia and steatosis of the liver. The excess weight facilitates the development of arthritis especially of

Table I. Consequences of obesity

Cardiovascular	- ischemic heart disease, hypertension, renal failure, stroke
Metabolic-Endocrine	- diabetes, gout, hypogonadism
Gastrointestinal	- hiatus hernia, steatosis, gall bladder disease
Inactivity	- cellulitis, deep venous thrombosis, pulmonary thromboembolism
Cancer	- GI and reproductive systems
Skeletal - arthritis	- especially back, hip and knees
Respiratory	- sleep apnea - hypoventilation
	respiratory muscle weakness
	reduced central responsiveness - CO_2, O_2
	Mechanics - reduced Vc, ERV, FRC, MVV
	increased closing volume
	chest wall compliance decreased
	high work of breathing
	Gas Exchange - $\dot{V}/\dot{Q}$ mismatch, hypoxia
	Cor Pulmonale
Death	

the lower back and knees which further impair mobility. The sedentary state favours the development of lower extremity cellulitis, deep vein thrombosis and pulmonary thromboembolism. There is also an excess risk for colorectal and gall bladder neoplasia, and cancers of the breast and reproductive organs.[7] These complications can impact on the ventilatory system secondarily.

The respiratory consequences of obesity have been noted since 360 BC when Dionysius died from choking on "his own fat"[21] Burwell called attention to the now classic description by Dickens of an obese 19 year old male who suffered from extreme hypersomnolence and the obesity hypoventilation syndrome.[3] The ventilatory consequences of obesity vary with the severity and extent of obesity, worsening with increasing mass. (Table I)

The obesity-hypoventilation syndrome is one of the most serious consequences of excess weight and while weight reduction helps, weight is not the only factor contributing to the syndrome.

Respiratory muscle weakness, possibly related to fatty infiltration of the respiratory muscles, is associated with hypercapnia, and contributes to the low maximum voluntary ventilation (MVV).[16] Central hyposensitivity may further aggravate hypoventilation since response to 5% CO_2 breathing is reduced.[20] In severe obesity, hypoxic drive may be reduced by 80%.[27] The combination of hypoxaemia and hypercarbia will lead to cor pulmonale.[13, 16, 20]

The excess chest wall mass reduces its outward recoil and lowers its compliance.[13,20] Lung compliance however, tends to be normal until obesity is more

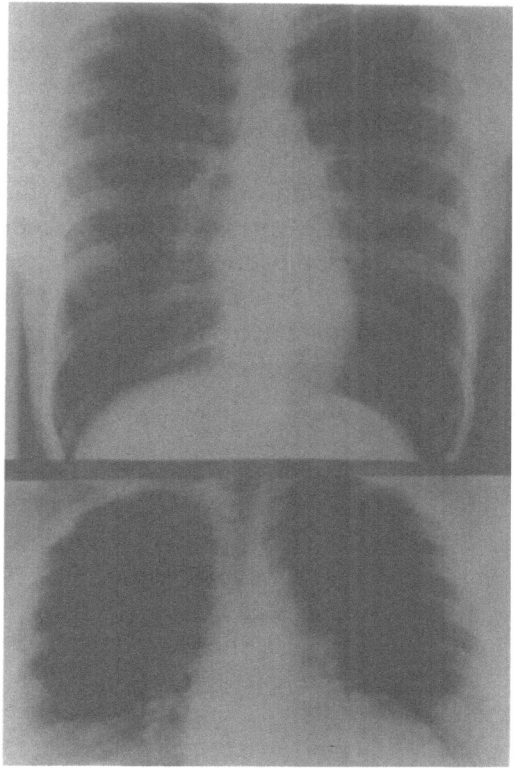

Fig. 1. Chest roentgenograms in an anorexic woman (upper panel) with 72% IBW; and an obese woman (lower panel) with 193% IBW. Both PA films are at maximal inspiration for the subject. The changes in lung volume can be noted from the degree of diaphragm descent and the distance between the ribs. No inference can be made from the x-rays regarding the mechanical properties of the lungs substending these thoraces.

severe. The increased mass of the abdominal wall and its contents result in reduced diaphragm descent, and thus a reduced vital capacity (VC), total lung capacity (TLC), functional residual capacity (FRC), and especially the expiratory reserve volume (ERV). Basal airway closure causes increasing closing volume (CV), and a fall in arterial oxygen tension (PaO_2). Persistent perfusion in these nonventilated paradiaphragmatic areas accounts for the ventilation/perfusion mismatch. Airways resistance is normal until subjects are more severely obese when dynamic compliance becomes frequency dependent even in non smokers.[6] Static lung compliance becomes reduced when polycythemia occurs.[15] All these adverse factors are aggravated by recumbency. They sum to increase the work of breathing, further adding to the increased CO_2 production and O_2 consumption already obligated by the increased body mass. Thus the efficiency of breathing is reduced and the cost of ventilation is high.

Exercise limitation in obesity can also be due to the increased metabolic rate and minute ventilation for the given work level. The excess metabolic cost of breathing stems from reduced chest wall compliance, the increased breathing frequency, and possibly from atelectasis due to dependent airway closure.[23]

Weight reduction is associated with reversing most of these abnormalities. However, it is not known whether the rapid weight loss attending extremely low caloric diets or intestinal bypass surgery have adverse effects on the respiratory system. My observations of some of these patients suggest a higher incidence of ventilator dependency and mortality due to pneumonia or after a surgical procedure. Much more work is needed to confirm this impression.

Consequences of Excessive Weight Loss

Weight loss may either be a cause or a consequence of ventilatory dysfunction. It may not be possible to determine which comes first. Emaciation and advanced emphysema have been associated with weight loss since the 19th century. Many studies have since described the association between increased mortality after the onset of weight loss in emphysematous patients, and the increased incidence of respiratory failure in those who had protein caloric malnutrition. This is in contrast to the patients with chronic bronchitis who only lost weight when they developed intercurrent bronchogenic carcinoma. Emphysema patients were more likely to have more severe reductions of FEV_1 and diffusing capacity (DLCO) which were significantly correlated with nutritional parameters.[24]

Human studies are limited by their complexity, lack of variable control and the variation in the rate of the development of malnutrition. Animal studies are limited by species differences, the short duration of diets imposed and the problem of extrapolating to humans. Much of world misery and disease is associated with inadequate nutrition, which in turn causes impairments which result in more disease and misery.

Nutritional deprivation assaults the respiratory system by decreasing ventilatory responses, reducing respiratory muscle function, impairing lung defense mechanisms, and disturbing lung development and repair. Each of these in turn affects the others. For example, a reduced ventilatory response can cause hypoxaemia and hypercarbia, which further hampers respiratory muscle performance which impairs cough and clearance resulting in infection and lung damage with defective repair. All these increase the work of breathing and when not compensable, cause respiratory failure and death.

In 1915, Benedict studied the physiological effects of diets restricted to 50/70% of the estimated requirements in normal volunteers and showed that a 10% weight loss was associated with a 15-20% reduction in basal metabolism which was

Table II. Hunger Disease. Studies by the Physicians of the Warsaw Ghetto[26]

Tiring climbing stairs
Tendency to catch cold, cough
Malabsorption - GI nutrient and blood loss
Osteomalacia - especially sternum and ribs
Reduced hormonal activity - thyroid, adrenal
Skeletal mass reduced
Increased infections due to leukopenia, delayed cell mediated immunity
Cardiovascular depression - bradycardia, hypotension, anemia, reduced O_2 and substrate transport
Emphysema in 13.5% (50/370 autopsies) where lungs were tympanitic, with low borders, low lung capacity, reduced capillary bed and minute ventilation

reflected in a mean decline of O_2 consumption by 18%, and minute ventilation by 12%.[2] Keys' studies showed a similar excess reduction of basal metabolism for the weight loss which was associated with a reduced efficiency of breathing with exercise. Refeeding revealed the reversibility of the decline of vital capacity, attributed to the effect of starvation on respiratory muscles. No compliance studies were done.[11]

The most unfortunate study of human starvation was performed by the heroic physicians of the Warsaw Ghetto during WW II.[26] Weight loss usually exceeded 50% of preinternment weight and progressed through several phases: first fat depletion, then withering and finally terminal cachexia. Basal metabolism was severely reduced, with impairment of glucose metabolism. They described easy fatiguability, tendency to cough and catch cold, malabsorption due to diarrhoea and GI bleeding, osteomalacia, including painful sternum and ribs, and becoming hypoendocrine. (Table II) The respiratory system was seriously affected by atony, low lung borders with tympany, reduced diaphragmatic excursion and lung capacity, decreases in minute ventilation due to both lower rate and depth of breathing, and loss of capillary bed. Reduced cardiac performance compromised the transport of both O_2 and substrates, worsened by anaemia.

R.Q. was 0.9 to 1.0 early, and fell to 0.7 later. There was an increased incidence of pulmonary infections, including bronchitis, pneumonia, and tuberculosis, associated with leucopenia. From detailed autopsy studies in 370 cases, 50 had changes like the "emphysema of old age", confirmed by microscopic studies in 5 of these.

Thirty-four of these were under 50 years of age. Although pathological definition and morphologic criteria were not the same as today, their studies support the concept that the ventilatory system, and the lung architecture can be profoundly affected by nutritional deprivation.

Studies of pulmonary function in ten anorexics by Cravetto et al. revealed a reduction of VC, lowest in the most malnourished female. The 5 with the lowest VC also had a reduced respiratory system compliance, thought to be related to the chest cage muscles. Expiratory flows and RV were normal.[5]

These limited studies appear to show that malnutrition affects the respiratory system predominantly by weakening the respiratory muscles, reducing metabolism and breathing efficiency when mild or moderate, and causes possibly irreversible structural changes when severe.

The effects of malnutrition on surface tension and tissue elastic properties, or on the metabolic functions of the lung in humans are unknown.

Animal studies are the only other resource to study the effect of malnutrition on the lung. It is probable that malnutrition alters the infrastructure of the lung by disrupting the balance between protein breakdown and resynthesis. (Table III) This dynamic matrix is in a continuum, where protases, elastases and collagenases remove damaged elastin, fibronectin and collagen. This proteolytic activity is counterbalanced by antiproteases such as alpha-1 antitrypsin.

In emphysema excesses of inflammatory cells release more proteolytic materials which are not adequately checked by antiproteases or antioxidants. Oxidizing agents are abundant in cigarette smoke, polluted air and products of phagocytic cells. Copper and iron are both necessary cofactors for functioning antioxidant enzyme systems such as superoxide dismutase and catalases. Selenium is a cofactor for glutathione reductase, and vitamins C and E are free radical scavengers. Thus, deficiency of these substances render antioxidant activity ineffective and facilitate lung injury. Monocytic cell recruitment to the lung, secretion and production of immunoglobulins (IgA), and cell mediated immunity are impaired by malnutrition. The defective clearance of bacteria and debris favours the destructive influences.[17,18,24]

Starvation reduces lung weight in proportion to body weight in young and old

Table III. Lung structural changes of severe starvation (rats)

Connective tissue loss
Remodel elastic fibers
Defective cross linking of elastin and collagen fibrils
Reduced proteins - RNA decreased greater than DNA
Reduced antiproteases, antioxidants, cofactors Fe, Cu, Se;
decreased monocytic recruitment to lung, synthesis of
interleukin I, cell mediated immunity and secretory
Immunoglobulin A

rats.[17] Depletion experiments are usually conducted over a few days to a few weeks with a fixed degree of diet restriction, which bears no resemblance to the human condition. To place into perspective the rat experiments as they may be related to man, I have estimated the approximate time man would have to be starved for the effects seen. The average life expectancy for a rat is 2 years and for man is 73 years. Thus, each day in a rat's life is equal to about 36 days in a man's life. Starvation rat experiments vary from 3 days to 3 weeks, which would be equivalent to about 4 months to 2 years for man, noting that constant diet restriction occurred only in the animals. Animal studies were also conducted to an arbitrary point of weight loss, usually 30% to 40% below control and this is severe, but not as severe as the Warsaw Ghetto experience.

In starvation protein synthesis in the lung is reduced with the reduction in RNA greater than DNA, indicating cell size is reduced more than cell number.[18] Collagen turnover is decreased as is the total amount of connective tissue. There is a remodelling of elastic fibers with shortening and disruption. The starvation induced reduction of lysyl oxidase alters the cross linkage of elastin and collagen fibers. Total protein deficiency results in lower connective tissue turnover and less structural emphysema than lower caloric intake which includes protein, suggesting that total protein deprivation reduces turnover less.[10] These biochemical and structural changes due to nutritional deprivation seriously alter lung mechanics, as shown by many and summarized by Sahebjami.[17] (Fig. 2)

Adult rats starved to 40% of control weight have a reduction of surfactant which results in increased surface elastic forces, shifting air volume-pressure deflation curves to the right. The granular pneumocyte, with less mitochondria, synthesizes less surfactant, reducing the storage form of surfactant the lamellar bodies . Saline V-P curves show that there is loss of tissue elastic recoil pressure, and this reduced elasticity shifts the V-P curve leftward, indicating an increased tissue compliance. Figure 2 shows the V-P curves, expressed in millimetres in A (upper panel), and percent of maximal lung volume in B, which are reduced in air, and increased in saline (lower panel). The changes are more severe at lower lung volumes. The difference in behaviour between air and saline suggest that the net change in man would depend on the severity of the systems affected. When the rats are refed, surfactant is rapidly repleted, restoring the air V-P curves. But the saline V-P curve remains shifted leftward, suggesting either incomplete restoration of elastin due to slower elastin repletion, or permanent irreversible damage.[18,19]

The scanning electron photomicrographs illustrate the effects of starvation on alveolar architecture in rat lungs, fixed at a transpulmonary pressure of 20 cm H20. (Fig. 3 A and B) The control rats (A) show uniform alveoli while the starved rats show enlarged air spaces and loss of uniformity. Further, the walls are thinner, with irregular interalveolar septae which are more effaced and contain more pores.[17]

Sahebjami also studied the effect of malnutrition on enzyme induced established emphysema, showing the synergistic effect of the two.[19] Six weeks of food deprivation to 1/3 of normal intake in normal adult rats (equivalent to 4 years in man) showed the highest mean linear intercept (Lm) in the starved rats with emphysema, indicating the larger interalveolar distances with the greatest reduction of internal

Lung mechanics in starvation

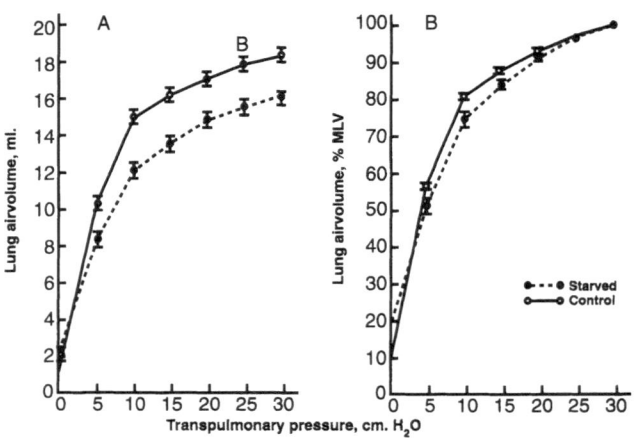

Air Pressure-Volume Curves

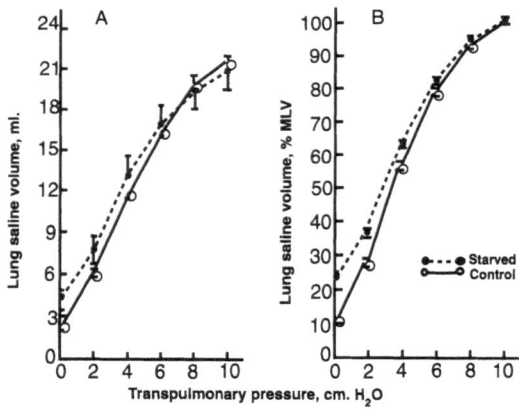

Saline Pressure-Volume Curves

Fig. 2 Static deflation air (upper panel) and saline (lower panel) in starved rats, expressed as millimeters (A) and percentage maximal lung volume (B). The starved curves are shifted right in air indicating a tendency to collapse, and left in saline indicating a decrease in tissue elasticity. From: Sahebjami H., Vasallo C.L., Wirman J.A.: Am. Rev. Resp. Dis. 1978; 117:77-83, with permission.

surface area (ISA) and the reduced surface active forces. (Fig. 4) The fact that no differences were observed between starved controls and fed emphysematous rats suggests that the effects are similar and that feeding did not reverse the abnormalities when emphysema is established.

Refeeding starved rats for 7 to 10 days is associated with partial recovery of Lm and ISA, a surge of DNA activity, restoration of phospholipids but incomplete restoration of connective tissue, elastin and total protein. This would account for the persistent leftward shift of the V-P curves.[18] (Table IV) The age when malnutrition occurs determines the effects of repletion: the earlier the starvation, the less the

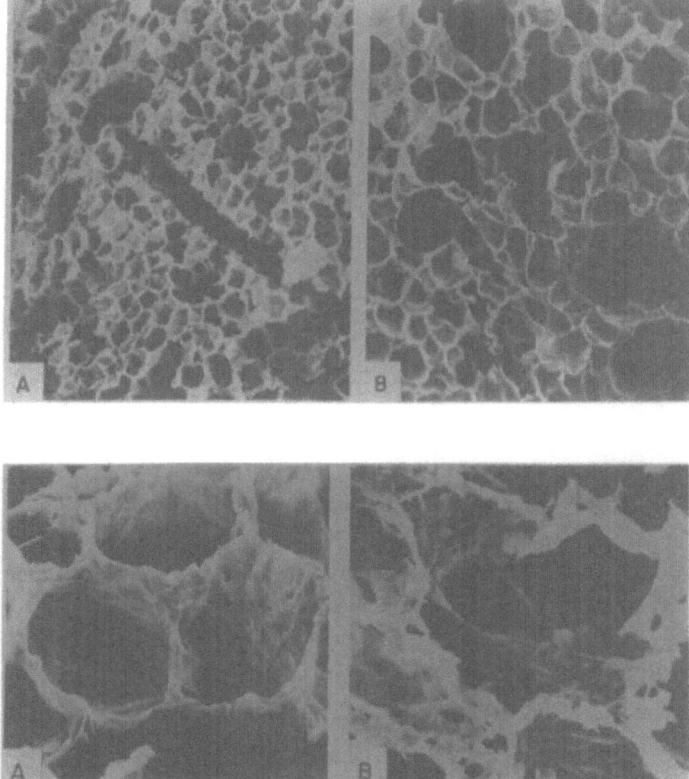

Fig. 3 Upper Panel. Scanning electronmicrographs of rat lungs fixed at a pressure of 20 cm H_2O in fed rats (A) and starved rats (B), showing the nonuniformity and enlarged air spaces consequent to starvation. (magnified x 150). Lower Panel. Scanning electronmicrographs of rat lungs fixed at a pressure of 20 cm H_2O of fed rats (A) and starved rats (B) showing that the air spaces in starved rats are larger, non uniform with more irregular and thinner interalveolar septae which contain more pores. (magnified x 800). From: Sahebjami H., Wirman J.A.: Am. Rev. Resp. Dis. 1981; 124.619-24, with permission

74

Table IV. Nutritional Repletion Effects on the Lung

Morphology	- partial recovery of mean linear intercept (Lm), internal surface area (ISA)
Biochemistry	- surfactant normalizes proteins with increased-burst DNA activity
Mechanics	- air V-P curve restored Saline V-P curves little changed
Growth	- depends on age malnourished - the earlier the starvation, the less the recovery; lung recovery more than the rest of the body
Unknown	- man with established emphysema and repletion

recovery. From electron micrographs and these lung mechanics studies, it is strongly suggested that elastic tissue changes could be permanent in man. The effects of repletion on these structural changes are not known in man.

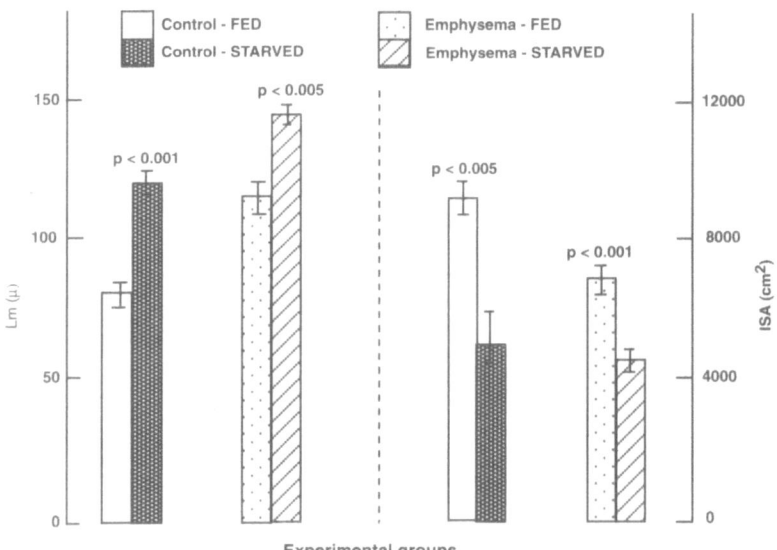

Fig. 4 Results of morphometric studies comparing control and enzyme induced emphysema for both fed and starved rats. Mean linear intercept (lm) is greater and internal surface area (ISA) is smaller in starved rats compared to fed rats for both control and emphysematous conditions. The lm is nearly the same for starved control and fed emphysematous rats. From: Sahebjami H., Vasallo C.L.: J. Appl. Physiol. 1980; 48:284-8, with permission

Effects of Repletion in Man

What then can we expect from nutritional intervention in patients with established emphysema? Studies to date have shown that caloric intake is not low in patients, but that despite high intake it still falls below caloric requirements for the energy actually expended. In this way patients with lung disease also differ from humans with starvation alone since basal metabolism is NOT decreased, and energy expended for breathing is so much higher. Lung and chest wall mechanics in emphysematous man have not been studied before and after nutritional repletion. Respiratory muscle strength, vital capacity and MVV may increase[9,25] or show no change.[1] Several studies have shown increases in FEV_1 with refeeding and one wonders whether airway inflammatory oedema may be lessened by better nutrition.[1,8] No one has reported on the effects of refeeding on infection rates, or the reduction of the rate of lung function decline over time.

Thus, both nutritional excesses and depletion impair ventilatory function in many ways. Studies are needed to determine the components reversible in damaged lungs, the types of diet needed to do this, whether antioxidants have any effect, how long these diets must be maintained and whether specific diet composition can alter the balance of the processes between lung injury and repair.

From personal experience, attention paid to all components in a patient, including improved diet, prevention of further lung injury by preventing infection, restoring normal sleep and improving respiratory muscle strength, can delay mortality and stabilize the course. Future studies are needed to determine what is possible to change and how change can be effectively accomplished.

References

1. Angelillo V.A., Bedi S., Durfee D., et al.: Effects on low and high carbohydrate feedings in ambulatory patients with chronic hypercapnia. Am. J. Med. 1955; 103: 88, 3-5
2. Benedict F.: *A study of prolonged fasting*. Washington, DC Carnegie Institute, (Publication 203)
3. Burwell C.S., Robin Ed., Whaley R.D., et al.: Extreme obesity associated with alveolar hypoventilation: A Pickwickian syndrome. Am. J. Med. 1956; 21: 811-1
4. Chiang S.T., Lee P.Y., Liu S.Y.: Pulmonary function in a typical case of Pickwickian syndrome. Respiration 1980; 39:105-13
5. Cravetto C., Carelli E., Cardillino G., et al.: Osservazione sulla funzione respiratoria nell'anoressia mentale. Minerva Med. 1966; 57:3498-501
6. Douglas F.G., Chang D.Y.: Influence of obesity on peripheral airways patency. J. Appl. Physiol. 1972; 33:559-63
7. Garfinkel L.: Overweight and cancer. Ann. Int. Med. 1985; (part 2) 103:1034-36
8. Goldstein S., Thomashaw B., Askanazi J.: Functional changes during nutritional repletion in patients with lung disease. Clin. Chest. Med. 1986; 7: 141-51
9. Kelly S.M., Rose A., Field S., et al.: Inspiratory muscle strength and body composition in patient receiving total parenteral nutrition therapy. Am. Rev. Resp. Dis. 1984; 130: 33-7

10. Kerr J.S., Riley D.J., Lanza-Jacoby S., et al.: Nutritional emphysema in the rat: influence of protein depletion and lung growth. Am. Rev. Resp. Dis. 1985; 131: 644-50

11. Keys A., Brozeh J., Henchel A., et al.: *The biology of human starvation*. Minneapolis, The University of Minnesota Press, 1950

12. Knowles G.B., Fairbarn M.S., Wiggs B., et al.: Dietary supplementation and respiratory muscle performance in patients with COPD. Chest 1988; 93: 977-83

13. Luce J.M.: Respiratory complications of obesity. Chest 1980; 78: 626-31

14. *1983 Metropolitan Height and Weight Tables*. Stat. Bull. Metrop. Life Insur. Co. 1984; 64: 2-9

15. Ray C.S., Sue D.Y., Bray G., et al.: Effects of obesity on respiratory function. Am. Rev. Resp. Dis. 1983; 128: 501-6

16. Rochester D.F., Enson Y.: Current concepts in the pathogenesis of the obesity-hypoventilation syndrome: mechanical and circulatory factors. Am. J. Med.1974; 57: 402-20

17. Sahebjami H.: Nutrition and the Pulmonary Parenchyma. Clin. Chest. Med. 1986; 7: 111-26

18. Sahebjami H., McGee J.: Influence of starvation and refeeding on lung biochemistry in rats. Am. Rev. Resp. Dis. 1982; 126: 483-7

19. Sahebjami H., Vasallo C.L.: Influence of starvation on enzyme induced emphysema. J. Appl. Physiol. 1980; 48: 284-8

20. Sharp J.T., Barrocas M., Chokroverty S.: The cardiorespiratory effects of obesity. Clin. Chest Med. 1980; 1: 103-18

21. Smith W. (ed): *A Dictionary of Greek and Roman Biography and Mythology*. Vol. II. London. John Murray 1880

22. Van Italie T.B.: Health implications of overweight and obesity in the United States. Ann. Int. Med. 1985; (pt. 2) 103: 983-988

23. Whipp B.J., Davis J.A.: The ventilatory stress of exercise and obesity: Exercise testing in the dyspneic patient. Am. Rev. Resp. Dis. 1986; 134: 872-7

24. Wilson D.O., Rogers R.M., Hoffman R.M.: Nutrition and chronic lung disease. Am. Rev. Resp. Dis. 1985; 132: 1347-65

25. Wilson D.O., Rogers R.M., Sander M.W.: Nutritional intervention in malnourished patients with emphysema. Am. Rev. Resp. Dis. 1986; 134: 872-7

26. Winick M. (ed): *Hunger disease: studies by the Jewish physicians in the Warsaw Ghetto*. New York, Wiley and Sons, 1979

27. Zwillich C.W., Sutton F.D., Pierson D., et al.: Decreased hypoxic ventilatory drive in obesity - hypoventilation syndrome. Am. J. Med. 1975; 59: 343-8

8. A Preliminary Report on the Effects of Malnutrition on Skeletal Muscle Composition in Chronic Obstructive Pulmonary Disease

E. Fiaccadori,[1] E. Coffrini,[1] N. Ronda,[1] G. Gonzi,[1] L. Bonandrini,[3] C. Fracchia,[2] C. Rampulla,[2] N. Ambrosino,[2] T. Montagna,[2] A. Borghetti[1]

1 Internal Medicine and Nephrology, University of Parma, Italy
2 Clinica del Lavoro Foundation, Institute of Care and Research Medical Center of Rehabilitation, Montescano, Pavia, Italy
3 Chair of Microsurgery, University of Pavia, Italy

Introduction

Several alterations of electrolyte metabolism have been demonstrated to impair respiratory muscle performance: respiratory muscle function improves after magnesium administration in hypomagnesiemic patients[1]; an increase in transdiaphragmatic pressure after the correction of hypophosphataemia has been found in hypophosphatemic patients[2]; low values of maximum inspiratory and expiratory pressure values were significantly increased after phosphate repletion in both hypophosphatemic surgical[3] and medical patients.[4]

Although mechanical disadvantage is thought to play a key role in the pathogenesis of respiratory muscle dysfunction in COPD patients, mineral deficiencies and electrolyte imbalances have also been indicated as important contributing factors to the respiratory muscle weakness demonstrated in this condition.[5-6] The presence of skeletal muscle composition derangements in COPD patients has been confirmed by several studies on peripheral skeletal muscles: magnesium and potassium content was decreased in the quadriceps femoris muscle of COPD patients with Acute Respiratory Failure, whereas sodium content was increased [7-8]; furthermore, phosphorus content was reduced in peripheral skeletal muscles of COPD patients, and a significant relationship between serum and muscle phosphorus values was demonstrated in the same patients.[9]

Nevertheless, data concerning electrolyte composition of respiratory muscles of COPD patients are still scanty, and only recently a reduced phosphorus content has been demonstrated in both respiratory and peripheral muscles of patients with severe COPD.[10] Moreover, although the alterations in electrolyte metabolism have been attributed, at least in part, to the presence of malnutrition, there are no data concerning possible differences in skeletal muscle electrolyte composition between COPD patients with malnutrition and COPD patients without malnutrition. The goals of our study were thus:

1. to evaluate the presence and severity of skeletal muscle electrolyte composition derangements in COPD patients, with particular emphasis on cationic and phosphorus content;
2. to evaluate the differences in electrolyte composition between respiratory and peripheral skeletal muscles of COPD patients;
3. to evaluate the differences in skeletal muscle electrolyte composition between COPD patients with normal nutritional status and COPD patients with malnutrition.

Patients and Methods

a. Patients

Twenty-two patients (19 male, 3 female, mean age 63 years ± 6SD) with hypercapnic-hypoxaemic COPD ($PaCO_2$ 57 mmHg ± 9, PaO_2 47±8) were studied. Patients were selected only if COPD was the primary diagnosis. A diagnosis of COPD had been made during previous hospitalizations or clinical visits and was based on positive history, clinical and radiological criteria, visits and standard measurements of pulmonary mechanics (FEV_1 values and FEV_1/FVC ratio less than 70% of predicted consciousness). No patient was diabetic or receiving either phosphorus supplementation or parenteral nutrition at the time of the study. Serum creatinine levels ranged from 0.8 to 1.2 mg%. Quadriceps femoris muscle needle biopsies were obtained in all COPD patients: in fourteen of the same patients surgical biopsies from the external intercostal muscles were also performed in local anaesthesia. Two groups of age and sex-matched subjects with normal hemogasanalysis and spirometry, as well as normal nutritional status, served as controls for the muscle biopsy study: 14 medical patients underwent quadriceps femoris needle biopsy in local anesthesia,[7-9] and 11 surgical patients hospitalized for elective thoracic or abdominal surgical procedures underwent external intercostal muscle surgical biopsies soon after general anaesthesia induction. In all COPD patients arterial hemogasanalysis, urinary and serum electrolytes, and creatinine were obtained on the day of the study. Nutritional status was evaluated in each control subject and COPD patient before the biopsy by measuring body weight (expressed as % of ideal body weight) serum albumin, transferrin, and total lymphocyte count, as previously described.[11]

Four criteria were used to define malnutrition: body weight, serum albumin levels less than 3.5 grams per decilitre, serum transferrin less than 200 mg%, total lymphocyte count less than two thousand per cubic millilitre. On these grounds, the COPD patient group was subdivided into two subgroups: COPD with malnutrition (10 patients) and COPD with normal nutritional status (12 patients). All subjects were informed about the nature and the possible risks of the study; written consent was obtained from each participant.

b. Methods

Arterial pH, PCO_2, PO_2, serum electrolytes (phosphorus, sodium, potassium, magnesium), and creatinine, were measured by routine analytical methods, as previously described.[9] Muscle specimens were obtained from the lateral portion of the quadriceps femoris, by the Bergstrom needle biopsy technique,[12] with a needle of 6 mm diameter (Bram, Parma, Italy).[8] Surgical biopsies from the external intercostal muscle of COPD patients were obtained, in local anaesthesia, in the 5th intercostal space, on the anterior axillary line by a thoracic surgeon. Muscle fragments were rapidly frozen in Freon 22 and stored in liquid nitrogen. They were then freeze-dried, extracted in petroleum ether, and muscle powder was obtained after separation of visible blood and connective tissues.[13] Muscle potassium, sodium, and magnesium content was measured after strong acid extraction by atomic spectrometry methods, as previously described.[7] Muscle phosphorus was obtained after strong acid extraction[14], according to the Chen method[15], which is based on the formation of a phosphomolybdate complex subsequently reduced by ascorbic acid and read with a mod. DU 65 Beckman spectrophotometer at 820 nm. Muscle electrolyte content was expressed as moles per kg of fat-free dry muscle weight (moles/kg of FFDM), i.e., the weight of muscle powder after freeze drying and ether extraction.

c. Statistics

Data are expressed as arithmetic mean ± standard deviation (SD). Statistical analysis was performed using the two-tailed Student's "t" test for unpaired data. Statistical significance was defined as a probability of type I error of less than 0.05. The personal computer packages "Works" (Microsoft Corp., Redmond, WA) and "Statpak" (Northwest Analytical, Portland, OR) were used for data storage, calculations, and statistics.

Discussion

Several studies in the last few years have provided increasing evidence that malnutrition is a common finding among patients with COPD:[11, 16-20,] it is in fact generally acknowledged that 30 to 40% of COPD patients are significantly undernourished, as judged by body weight loss and anthropometric estimates of body fat and muscle mass. Malnutrition in patients with COPD is associated with increased frequency and extent of hospitalization, increased morbidity, development of respiratory failure and mortality.[21-24] Factors responsible for malnutrition and weight loss in COPD patients are not fully defined and a number of theories has been proposed. Several factors are probably involved in producing the final effect of weight loss: inadequate dietary intake of nutrients, hypermetabolism, limited cardiac output, depression etc.[25] Consequences of malnutrition are manifold, and several important derangements have been demonstrated in lung repair processes, surfactant production, defence mechanisms, control of ventilation and respiratory muscle structure and function.[25] In particular, malnutrition is commonly associated with a generalized lean body mass depletion which also involves specific wasting of respiratory muscles. This effect on lean body mass has been demonstrated both in experimental studies and in humans.[26-30] In course of malnutrition diaphragm muscle changed in parallel with body weight and both the area and the thickness of diaphragm muscle changed in parallel with muscle mass, so that about half of the increase or the decrease in mass was due to a corresponding increase or decrease in a highly significant linear relationship between either diaphragm muscle mass or thickness and body weight was found. Thus diaphragm muscle mass changes according to the nutritional status in the same way as that of the other body muscles.

In patients with COPD diaphragm muscle mass and thickness vary with body weight in the same way as in a patient without COPD.[30-31] Mechanisms of muscle wasting in COPD are poorly understood. Yet, little research has been devoted to this problem despite the fact that muscle wasting is commonly thought to be associated with a poor prognosis in COPD patients. In a recent study on malnourished patients with emphysema,[32] whole-body leucine flux was normal, whole-body leucine oxidation was increased, and whole-body protein synthesis was depressed. These results indicate that the predominant mechanism of muscle wasting in emphysema is a fall in muscle protein synthesis, which is accompanied by a fall in overall whole-body protein turnover. Among the possible mechanisms for depressed protein synthesis rate in COPD, hypoxia, energy deficit due to the increased metabolic cost of breathing, reduced intake, absorption and utilization of nutrients, relative immobility leading to disuse atrophy, and the effects of drug therapy such as corticosteroids have been postulated.[32]

Malnutrition results in a deterioration of both peripheral[33] and respiratory muscle function[34-35] that exceeds the loss of muscle mass. Undernourished subjects, with body weight of 71% of ideal, had a respiratory muscle strength that was only 37% of normal; both inspiratory and expiratory muscles were affected to the same extent. Also in malnourished patients undergoing nutritional repletion muscle function improves within two weeks, that is, long before lean body mass is reduced out of proportion to the loss of muscle mass, and that the respiratory muscle weakness cannot be explained by the undernutrition-linked decrease in muscle function which may be due not only to a reduction in muscle mass, but also to alterations in the remaining muscle cells, i.e., to derangements in muscle composition.[5]

In our study we have demonstrated that important derangements in skeletal muscle composition characterize patients with severe COPD. These abnormalities occur to the same extent in both resting limb muscles and active respiratory muscles, but are more severe when a malnutrition coexists with COPD. Our results are thus fitting in with the hypothesis that the adverse effects of malnutrition on muscle function cannot be attributed to a simple loss of body mass.

Electrolyte metabolism derangements are among the most common causes of myopathy: phosphorus, potassium, and magnesium play in fact a major role in numerous essential biochemical processes of the skeletal muscle cell and are critical to many aspects of muscle cell integrity, muscle contraction and central nervous system function (Fig. 1-4).[37]

Clinical features of hypophosphataemia and of hypomagnesaemia include

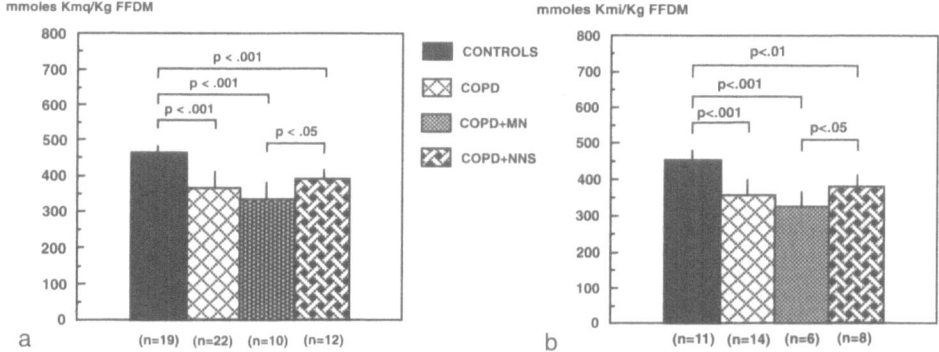

Fig. 1a Quadriceps femoral muscle Potassium (Kmq) in control subjects and COPD patients. b External Intercostal muscle Potassium (Kmi) in control subjects and COPD patients.

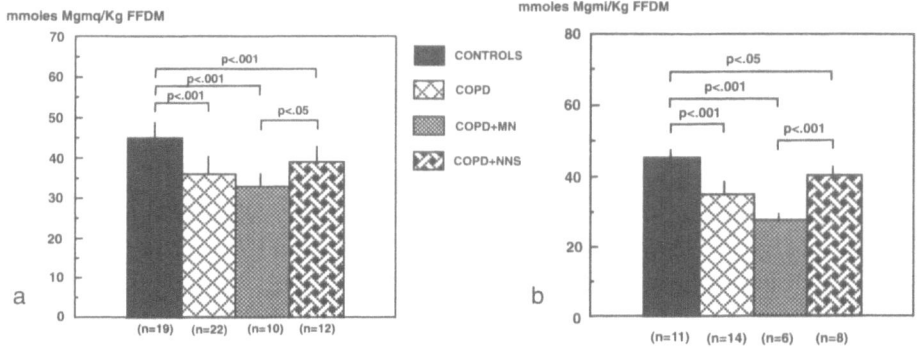

Fig. 2a Quadriceps femoral muscle Magnesium (Mgmq) in control subjects and COPD patients. b External Intercostal muscle Magnesium (Mgmi) in control subjects and COPD patients.

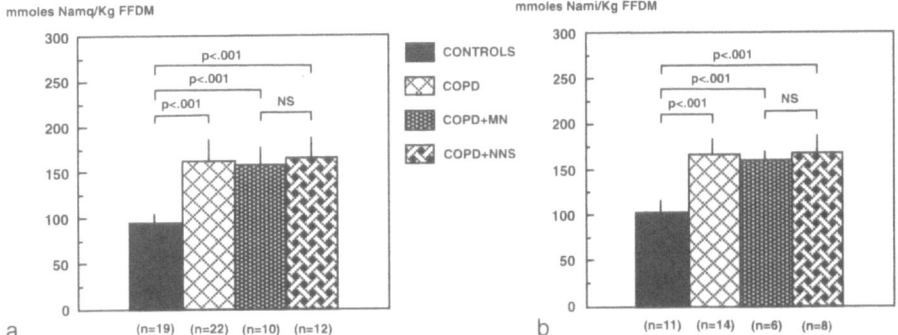

Fig. 3a Quadriceps femoral muscle Sodium (Namq) in control subjects and COPD patients. b External Intercostal muscle Sodium (Nami) in control subjects and COPD patients.

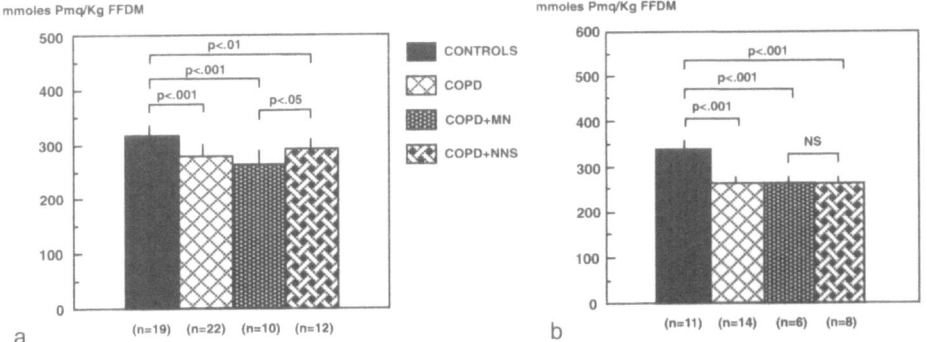

Fig.4a Quadriceps femoral muscle Phosphorus (Pmq) in control subjects and COPD patients. b External Intercostal muscle Phosphorus (Pmi) in control subjects and COPD patients.

important neuromuscular abnormalities; hypokalaemia and/or potassium depletion is associated with a subnormal resting membrane potential, high intracellular sodium levels, impaired blood flow during exercise, muscle weakness and rhabdomyolisis.[37]

Molecular mechanisms mediating skeletal muscle myopathy of electrolyte disorders are not completely defined. It is likely that these abnormalities are ultimately able to induce an overall compromise of cell energy production, storage, transport, and utilization in the form of ATP and phosphocreatine.[37]

The clinical importance of these abnormalities in COPD patients is so far unknown. In fact, although markedly reduced levels of ATP and phosphocreatine have been observed in the skeletal muscle of COPD patients[38-40], the extent to which electrolyte metabolism derangements are responsible for both muscle energy metabolism alterations and muscle functional impairment is not well defined. Moreover, it must be stressed that, besides malnutrition, many other factors are able to influence negatively skeletal muscle composition in COPD patients. Skeletal muscle magnesium was in fact lower in COPD patients who had been on therapy with aminoglycoside antibiotics[8], which are known to induce renal magnesium wasting. The prevalence of hypophosphataemia was higher among COPD patients taking one or more drugs commonly used in COPD therapy and known as negatively affecting renal phosphate handling: e.g. xanthine-derivatives, corticosteroids, diuretics, and β_2 adrenergic bronchodilators.[9] Finally, it must be emphasized that determinants of respiratory muscle function are numerous[5-6] and many factors have been demonstrated to impair respiratory muscle function, such as acidosis[41], derangements in respiratory muscle length-tension relationship, alterations of blood flow and oxygen delivery to respiratory muscles.[42-43]

In conclusion, in the COPD patients we have studied it is evident that:

a. skeletal muscle of patients with severe COPD is characterized by important electrolyte content derangements;
b. these derangements equally affect both peripheral and respiratory muscle;
c. such alterations in muscle composition can be present regardless of nutritional status, though they are worse in the presence of nutritional depletion.

Future research should enable us to obtain more definitive data about these open problems:

1. Relationships between severity of COPD and alterations in respiratory muscle composition;
2. Relationships between respiratory muscle composition and function in COPD patients;
3. Effects of nutritional repletion on respiratory muscle composition and function in malnourished COPD patients.

References

1. Molloy D.W., Dhingher S., Solven F., Wilson A., McCarthy: Hypomagnesiemia and respiratory muscle power. Am. Rev. Resp. Dis. 1984; 129:497-498
2. Aubier M., Murciano D., Lecocguic Y., Viires N., Jacquens Y., Squara P.: Effect of hypophosphatemia on diaphragmatic contractility in patients with acute respiratory failure. New Engl. J. Med. 1985; 313:420-424
3. Gravelyn T.P., Brophy N., Siegert C., Peters-Golden M.: Hypophosphatemia-associated respiratory muscle weakness in a general inpatient population. Am. J. Med. 1988; 84:870-976
4. Horst H.M., Twyman D.L., Force S., Bivins B.A.: Effect of hypophosphatemia on inspiratory force in patients receiving total parenteral nutrition. J. Parent. Ent. Nutr. 1986; 10 (suppl.) 17S
5. Rochester D.F., Arora N.S.: Respiratory failure. Med. Clin. North Am. 1983; 67:573-598
6. Grassino A.: Determinants of respiratory muscle failure. Am. Rev. Resp. Dis. 1986; 134:1091-1093
7. Fiaccadori E., Del Canale S., Arduini U., Coffrini E., Antonucci C., Borghetti A.: Intracellular

acid-base and electrolyte metabolism in skeletal muscle of patients with chronic obstructive lung disease and respiratory failure. Clin. Sci. 1986; 71:703-712

8. Fiaccadori E., Del Canale S., Coffrini E., Vitali P., Guariglia A., Borghetti A.: Muscle and serum magnesium in pulmonary intensive care unit patients. Crit. Care Med. 1988; 16:751-760

9. Fiaccadori E., Coffrini E., Ronda N., Vezzani A., Cacciani G.C., Fracchia C., Rampulla C., Borghetti A.: Hypophosphatemia in course of chronic obstructive pulmonary disease: prevalence, mechanisms and relationships with skeletal muscle phosphorus content. Chest 1990; 97: 857-868

10. Fiaccadori E., Coffrini E., Vitali P., Ronda N., Guariglia A., Fracchia C., Rampulla C., Ambrosino N., Montagna T., Zocchi L., Bonandrini L., Borghetti A.: Low phosphorus content in axial and respiratory muscles of patients with chronic obstructive pulmonary disease. Am. Rev. Resp. Dis. 1989; 139:A166

11. Fiaccadori E., Del Canale S., Coffrini E., Vitali P., Antonucci C., Cacciani G.C., Mazzola I., Guariglia A.: Hypercapnic-hypoxemic chronic obstructive pulmonary disease (COPD). influence of severity of COPD on nutritional status. Am. J. Nutr. 1988; 48:680-685

12. Bergstrom J.: Muscle electrolytes in man. Scand. J. Clin. Lab. Invest. 1962; 14:1-110

13. Harris R.C., Hultman E., Nordesjo L.O.: Glycogen, glycolitic intermediates and high-energy phosphates in biopsy samples of musculum quadriceps femoris of man at rest. Methods and variance of values. Scand. J. Lab. Invest. 1974; 33:109-117

14. Montanari A., Borghi L., Curti A., Mergoni M., Sani E., Elia G.: Skeletal muscle cell abnormalities in acute hypophosphatemia during parenteral nutrition. Min. Electr. Metab. 1984; 10:52-57

15. Chen P.S., Toribara T.Y., Warner H.: Microdetermination of phosphorus. Analyt. Chem. 1956; 28:1756-1758

16. Hunter A.M.B., Carey M.A., Larsh H.W.: The nutritional status of patients with chronic obstructive pulmonary disease. Am. Rev. Resp. Dis. 1981; 124:376-381

17. Openbrier D.R., Irwin M.M., Rogers R.M.: Nutritional status and lung function in patients with emphysema and chronic bronchitis. Chest 1982; 87:17-22

18. Driver A.G., LeBrun M.: Iatrogenic malnutrition in patients receiving ventilatory support. JAMA 1980; 244:2195-2196

19. Driver A.G., McAlevy M.T., Smith J.L.: Nutritional assessment of patients with chronic obstructive pulmonary disease and acute respiratory failure. Chest 1982; 82:568-571

20. Keim L.K., Luby M.H., Braun S.R., Martin A.M., Dixon R.M.: Dietary evaluation of outpatients with chronic obstructive pulmonary disease. J. Am. Diet. Ass. 1986; 86:902-906

21. Boush S.F., Adhukari P.K., Sakamoto A.: Factors affecting prognosis in emphysema. Br. J. Dis. Chest. 1964; 45:402-411

22. Traverse G:A:, Cline M:G:, Burrows B.: Predictors of mortality in chronic obstructive pulmonary disease. Am. Rev. Resp. Dis. 1979; 119:895-902

23. Butler C.: Diaphragmatic changes in emphysema. Am. Rev. Resp. Dis. 1976; 114:155-159

24. Braun S.R., Dixon R.M., Keim N.L.: Predictive value of nutritional assessment factors in COPD. Chest 1984; 85:353-357

25. Wilson D.O., Rogers R.R., Openbrier D.: Nutritional aspects of chronic obstructive pulmonary disease. Clin. Chest Med. 1986; 7:643-656

26. Rowe R.W.D.: Effects of low nutrition on size of striated muscle fibers in the mouse. J. Exp. Zool. 1968; 167:353-358

27. Li J.B., Goldberg A.L.: Effects of food deprivation on protein synthesis and degradation in rat skeletal muscles. Am. J. Physiol. 1979; 236:E222-E228

28. Kelsen S.G., Ference M., Kappoor S.: The effects of prolonged undernutrition on the structure and functional of the diaphragm. J. Appl. Physiol. 1985; 58:1354-1359

29. Arora N.S., Rochester D.F.: Effect of body weight and muscularity on human diaphragm muscle, thickness and area. J. Appl. Physiol. 1982; 52:64-70

30. Arora N.S., Rochester D.F.: Effect of chronic airflow limitation (CAL) on sternocleidomastoid muscle thickness. Chest 1984; 85:56S

31. Rochester D.F.: Malnutrition and the respiratory muscles. Clin. Chest. Med. 1986; 7:91-99

32. Morrison W.L., Gibson J.N.A., Scrimgeour C., Rennie M.J.: Muscle wasting in emphysema. Clin. Sci. 1988; 75:415-420

33. Lopes J., McR Russell D., Whitwell J., Jee jeebhoy K.: Skeletal muscle function in malnutrition. Am. J. Clin. Nutr. 1982; 36:602-610

34. Arora N.S., Rochester D.F.: Respiratory muscle strength and maximal voluntary ventilation in undernourished patients. Am. Rev. Resp. Dis. 1982; 126:5-8

35. Fraser I.M., Russell D.M., Whittaker S.: Skeletal and diaphragmatic muscle function in malnourished COPD patients. Am. Rev. Resp. Dis. 1984; 129:A269

36. Russell D.M., Walker P.M., Leiter L.A.: Metabolic and structural changes in skeletal muscles during hypocaloric dieting. Am. J. Clin. Nutr. 1984; 39:503-513

37. Knochel J.P.: Neuromuscular manifestations of electrolyte disorders. Am. J. Med. 1982; 72:521-535

38. Gertz I., Hedenstierna G., Hellers G., Wahren J.: Muscle metabolism in patients with chronic obstructive lung disease and acute respiratory failure. Clin. Sci. Mol. Med. 1977; 52:395-403

39. Campbell J.A., Hughes R.L., Sahgal V., Frederiksen J., Shields T.W.: Alterations in intercostal muscle morphology and biochemistry in patients with obstructive lung disease. Am. Rev. Resp. Dis. 1980; 122:679-686

40. Fiaccadori E., Del Canale S., Vitali P., Coffrini E., Ronda N., Guariglia A.: Skeletal muscle energetics, acid-base equilibrium and lactate metabolism in patients with severe hypercapnia and hypoxemia. Chest 1987; 92:883-887

41. Rochester D.F.: The diaphragm: contractile properties and fatigue. J. Clin. Invest. 1985; 75:1397-1402

42. Juan G., Calverley P., Talamo C., Schnader J., Roussos C.: Effect of carbon dioxide on diaphragmatic function in human beings. New Engl. J. Med. 1984; 310:874-879

43. Roussos C.: Function and fatigue of respiratory muscles. Chest 1985; 88S:124S-132S

9. Malnutrition and Lung Host Defenses: Implications for the Pathogenesis and Prevention of Pneumonia

M.S. NIEDERMAN

Medical and Respiratory Intensive Care Unit, Winthrop-University Hospital, SUNY, Stony Brook, New York, USA

The role of nutritional impairment in promoting respiratory infection has been recognized for many years, but only recently have some of the mechanisms of malnutrition-induced respiratory host defense failure been clarified. Malnutrition has multiple influences on the lung's ability to resist invading pathogens and these alterations in lung defenses lead to three clinical consequences: an increased incidence of lung infections, a more virulent and prolonged bout of infection than is usually seen, and a more subtle clinical presentation of pneumonia than is typical of well-nourished individuals.[10]

In this review, the role of malnutrition in respiratory infection is discussed. First, the data relating to an increased incidence, severity and altered clinical presentation of pneumonia in the malnourished individual are presented. Then, the specific host defense impairments that have been observed in a variety of human and animal malnutrition states are reviewed.

Finally, we examine the difficulties in nutritional repletion as they relate to lung infection. Although it is easy to demonstrate that malnutrition promotes lung infection, it is much more difficult to show that nutritional therapy can reduce infection risk or that it can be safely given without further promoting the development of pneumonia.

Recent investigations have shown that the intestinal tract can promote pneumonia by serving as a source of bacteria for the lung and by providing organisms that can undergo translocation across the intestinal wall, thereby promoting systemic sepsis.[2,7,12,21]

Unless these factors are considered in nutritional replacement protocols, the incidence of lung infection and systemic sepsis may be increased by the very therapy intended to help the patient suffering from malnutrition.

The Influence of Malnutrition on the Incidence and Clinical Features of Pneumonia

Malnutrition and the incidence of lung infection

Children. In developing countries, the incidence of bronchopneumonia and viral lung infections is increased in children who are malnourished. In a study of 75 malnourished infants in Kampala, infection of multiple types was common, with 12 infants developing pneumonia, including tuberculosis and possibly gram-negative infection.[20]

In Bangladesh, similar findings have been reported, with nearly half of severely malnourished children having pneumonia, including tuberculosis in more than 10 percent.[3]

Viral infections, including measles and viral bronchitis may also be more common in malnourished children.[10]

Adults. In studies of survivors of ghetto conditions and prisoner of war camps in World War II, the incidence of pneumonia was increased in adults who were severely malnourished. In the Warsaw ghetto, tuberculosis was common, and in prisoners of war who were poorly fed, tuberculosis was seen with a higher than expected frequency.[10] In this latter population, common respiratory infections were not present with an increased incidence. In a variety of medical and surgical patients, the finding of abnormal nutritional parameters is common.

Driver found that in patients with respiratory failure and chronic obstructive lung disease often had abnormalities in triceps skinfold thickness, serum transferrin and total lymphocyte count.[6] In that study, 56% of respiratory failure patients were below 80% of ideal body weight.

This situation may be compounded by medical therapy because patients treated with mechanical ventilation are often underfed. One study observed that on average, among 26 patients receiving mechanical ventilation, daily caloric intake was 390 calories below basal metabolic needs, even though 71% of these patients had a serum albumin below 3.2 g/dl.[5]

In a surgical population, Mullen observed that nutritional impairment was not only common, but was also related to an increased risk of post-operative infection.[13] For a group of 64 elective surgical patients, 16 nutritional and immunologic parameters were measured, and 97% of the patients had at least one abnormality, while 35% had three or more abnormal findings.[13] Of the 64 patients, 15 developed a serious post-operative complication, including 9 infections, 3 of which were pneumonia.

These complications were more common in patients with the lowest serum albumin, transferrin, and delayed hypersensitivity testing.[13]

Malnutrition and the severity and clinical presentation of pneumonia

In both adults and children, malnutrition may favour a more serious course or even a fatal outcome of respiratory infection. In the Warsaw ghetto experience, tuberculosis was more severe in malnourished children, and pneumonia often presented almost asymptomatically.[10] Both of these observations reflect an impairment in host defense function, which must be intact to ensure the best outcome from infection, and the most appropriate inflammatory (and thus symptomatic) response to invading pathogens. In children in Bangladesh, mortality rates from pneumonia are higher in the presence of malnutrition.[3,10]

In a recent study of 80 patients with suppurative lung disease requiring surgical therapy, nutritional status was found to be an important determinant of outcome.[19] In this group, 64 patients had purulent empyema, 11 had tuberculous empyema or pleural effusion, 2 had lung abscess, and 3 had other infections. Fatal outcome was best correlated with a low serum albumin, but mortality was also affected by the presence of reduced delayed hypersensitivity testing.[19] Studies such as these do demonstrate the increased frequency and poor outcome of respiratory infection in those with malnutrition, but they do not tell us whether nutritional therapy can be employed to reverse this trend.

Malnutrition and Specific Respiratory Host Defenses

Normal Host Defenses

Once microorganisms enter the respiratory tract, they encounter a host defense system that has the capacity to repel them, at every anatomic site. Whether pneumonia develops, is a function of the size of the bacterial inoculum, the virulence of the specific organism, and the ability of the host to resist infection by virtue of having an intact defense system.

The oropharynx is usually free of gram-negative bacilli, in normals, due to the cleansing action of secretions, salivary proteases (lysozyme, lactoferrin, and lactoperoxidase), and local IgA antibody.[15] The oropharyngeal epithelium resists gram-negative colonization by cellular desquamation, and an intrinsically poor ability of oral mucosal cells, to adhere to this type of bacteria. With illness, mucosal cells, of both the upper and lower airway, can bind more gram-negative bacteria than normal, and consequently the airway can become colonized. For this to occur, cells expose more "receptors" that bind to "adhesins" on appropriately-equipped bacteria.

The lower respiratory tract has an even more complex host defense system that keeps this site sterile in normal individuals.[15] To colonize and infect the lower airway, bacteria must first navigate the epiglottis and vocal cords, then pass through the physical barriers of the tracheobronchial tree. These physical defenses include cough, bronchoconstriction, airway angulation and the mucociliary escalator.

Beyond these barriers are the immunologic defenses which can be organism-specific or non-specific. These include bronchus associated lymphoid tissue, phagocytic cells (polymorphonuclear leukocytes and alveolar macrophages), humoral immunity (IgA antibody, IgG antibody, and complement), and cell-mediated immunity with T lymphocytes. Also in the lower airway, as in the oropharynx, bacterial adherence is necessary for gram-negative bacteria to colonize the mucosa.[14]

Certain bacteria are handled by the lower airway host defense system in very specific ways. For example, *Staphylococcus aureus* is primarily eliminated by resident alveolar macrophages, while certain gram-negative bacteria require the recruitment of neutrophils in order to be eliminated.[10] Another organism, *Listeria monocytogenes*, which has been the subject of some studies of the effect of malnutrition, requires alveolar macrophages to be recruited to the lung in large numbers, because resident macrophages alone cannot effectively remove this organism.[10]

Host defenses with malnutrition

The increased incidence of lung infection that has been seen in patients with malnutrition, may be related to specific impairments in the respiratory tract host defense system. Malnutrition has been studied in a variety of animal models and human disease conditions. In these studies, there is no uniformity regarding the type of nutritional deficit being studied.

Some reports evaluate adult hospitalized patients with protein-calorie malnutrition, others examine children with protein-calorie deficits, while still others evaluate global indices of nutritional condition. Animal studies have evaluated acute starvation, chronic protein-calorie malnutrition, or combined food and water deprivation, in both mature animals and neonates. Combining all of these studies, the reported effects of malnutrition on host defenses are multiple and are summarized in Table I, and enumerated in greater detail below.

Table I. Lung host defenses with malnutrition

... Increased buccal and tracheal cell bacterial adherence: airway colonization
... Impaired alveolar macrophage phagocytosis (but not killing)
... Impaired alveolar macrophage recruitment to lung (but normal chemotaxis)
... Reduced alveolar macrophage production of LT B4, a PMN chemoattractant
... Abnormal neutrophil mobility (but normal phagocytosis and killing)
... Impaired cell-mediated immunity: abnormal lymphocyte function
... Reduced respiratory tract secretory IgA
... Complement deficiency

Increased respiratory cell adherence for gram-negative bacteria

Animal studies have shown that food and water deprivation (starvation) can increase buccal cell receptor density for *Pseudomonas aeruginosa*, thus serving as a mechanism for malnutrition to promote oropharyngeal colonization.[9] Higuchi et al. reported that rats exposed to starvation, developed a serial increase in the ability of their buccal epithelial cells to bind bacterian that paralleled loss of body weight. This cellular change translated into an increased likelihood for the oropharynx to become colonized, when challenged with *Pseudomonas aeruginosa*. The changes in cellular adherence were rapid, and developed after only three days of food and water deprivation, and returned to normal levels after two days of refeeding.[9]

In humans, colonization of the lower respiratory tract has also been related to malnutrition, possibly mediated by increases in tracheal cell adherence for gram-negative bacteria. We have studied patients treated with both chronic tracheostomy and endotracheal intubation and have related their colonization status to a nutritional assessment.[18]

Nutrition was evaluated with a multifactorial index, the Prognostic Nutritional Index (PNI), that incorporated measurements of triceps skinfold thickness, serum albumin, serum transferrin, and total lymphocyte count. As nutritional impairment increased, so did the numerical value for the PNI. Among 15 patients with chronic tracheostomy, the PNI was significantly higher for those colonized by *Pseudomonas* species than for those without this finding.[18] When tracheal cell adherence was measured in these patients, colonized patients had a higher mean adherence for *Pseudomonas aeruginosa* than non-colonized patients, and the absolute degree of tracheal cell adherence was directly related to the PNI for each patient. Thus, patients with the worst nutritional condition (highest PNI), tended to have the greatest ability of their airway epithelial cells to bind bacteria, indicating that malnutrition may have acted to cause airway colonization by increasing the number of epithelial cell receptors available for bacterial binding. One patient was studied serially, after vigorous enteral nutritional therapy for 10 weeks, and this individual showed a marked decline in his PNI, along with a similar decline in tracheal cell adherence and a loss of *Pseudomonas aeruginosa* from his sputum.[18]

A similar influence of nutritional status was observed in mechanically ventilated patients. In a study of 14 individuals treated with mechanical ventilation for a least one week, sputum was cultured every third day for a minimum of three times, and *Pseudomonas* species were found in 41 percent of all lower respiratory tract cultures.[16] However, among 6 patients with the highest PNI's, more than 60% of the cultures harbored *Pseudomonas* species, in contrast to only 25 percent of cultures with this finding among the 8 patients with a better nutritional status and a lower PNI.[16] Also in that study, malnutrition appeared to synergize with the effects of colonization of the trachea by non-*Pseudomonas* bacteria, to make the trachea more receptive for subsequent colonization by *Pseudomonas* species.

Alveolar macrophage function.

A variety of animal studies has evaluated this effector cell and has shown that different types of nutritional insults can lead to a loss of normal macrophage function. Reduced phagocytosis of gram-negative bacteria by resident alveolar macrophages, and reduced recruitment of these cells to the lung, when challenged by pathogens such as *Listeria monocytogenes*, have been observed.[10,11,23]

Green and Kass performed one of the earliest studies in this area by exposing mice to acute starvation, with the removal of food but not water, for 24 to 48 hours.[8] Animals were then exposed to aerosolized *Staphylococcus aureus* and whole lung clearance of the organism was measured. The data demonstrated that bacterial clearance was reduced, in direct proportion to the amount of weight lost by each animal. Although specific cellular functions were not evaluated, the methodology and type of bacteria used, implied that the reduced ability to clear the challenging organism was the result of macrophage dysfunction.

Shennib et al. also studied the effects of acute food (but not water) starvation, but used rats as the experimental subjects.[23] In that study, animals were killed after one week of deprivation and their alveolar macrophages were lavaged from the lung for further evaluation. Nearly 40 percent of the animals either died or developed serious pulmonary infection. The recovered macrophages had a reduced ability to phagocytose *Pseudomonas aeruginosa*, but bactericidal function was intact for any organisms that were ingested. In one study group, animals were starved and then refed, but macrophage function did not normalize until 3 weeks after resumption of a normal diet. This finding suggested that the effect of starvation on the macrophages may have been an irreversible one, and that new cells needed to be synthesized in order for normal function to return. This was in contrast to the finding of starvation-induced lymphopenia, which rapidly reversed within one week of refeeding.[23]

The effects of chronic protein-calorie malnutrition differ from the findings in animals exposed to acute starvation, and some elegant investigations with this model have been done using neonatal rats.[10,11] Martin and associates fed animals with a protein and calorie deficient diet for 4 weeks, and found that they developed a syndrome similar to human kwashiorkor, with marked loss of body weight.[11] When the animals were exposed to aerosol challenges of *Staphylococcus aureus* and *Pseudomonas aeruginosa*, whole lung clearance was intact, suggesting that the function of resident phagocytes and the ability to recruit neutrophils to the lung, were preserved. However, these animals could not adequately remove *Listeria monocytogenes*, when this organism was given as an aerosol challenge, and many of them died as a result of this exposure. The failure to clear the *Listeria* may have related to an inability of the malnourished animals to recruit alveolar macrophages to the lung after bacterial exposure. Although alveolar macrophages from the chronically malnourished animals had intact responses to chemotactic factors and normal bactericidal function, these cells did not accumulate in the increased numbers, seen in control animals, that were needed to eliminate the *Listeria*. Along with this deficit in macrophage function, the T-cell response to mitogens was reduced.[11] More recently, these investigators have found that alveolar macrophages from chronically malnourished infant rats have a reduced ability to produce leukotriene B4, a neutrophil chemoattractant.[10]

Polymorphonuclear (neutrophil) cell function

Although protein-calorie malnutrition has a mild effect on neutrophil function, which usually has little clinical implication, some affected children are unable to develop suppurative lesions in the presence of soft tissue infection, developing necrotic lesions instead.[10]

Anderson et al. evaluated 37 children with protein-calorie malnutrition and found abnormalities in the early phase migration of blood neutrophils, in conjunction with evidence of in vivo activation of these cells which may have prevented them from responding appropriately to chemotactic stimuli.[1] Serum pre-albumin levels were the best correlate of this abnormal neutrophil migration. With nutrition therapy, the abnormalities improved after 3 to 4 weeks, but clinical infection did not relate to abnormal neutrophil function. Other investigations have found that the microbicidal and phagocytic function of neutrophils is not affected by protein deficiency.[10]

Cell mediated immunity: lymphocyte function

In many studies of humans and animals, abnormalities in delayed type hypersensitivity and lymphocyte count have been found. In fact the extent of these expected aberrations is incorporated into nutritional assessment formulas such as the Prognostic Nutritional Index.[18] Skin test anergy is a common finding in patients with protein-calorie malnutrition, as is a reduction in total lymphocyte count and T cell number.[4] In malnourished children, there is a reduction in the number of T lymphocytes in peripheral blood, along with a normal number of B lymphocytes, and increased number of "null", or undifferentiated, cells. This finding has suggested that malnutrition may interfere with the differentiation of lymphocytes, possibly by causing a reduction in thymic hormone.[4] The lymphocyte response to mitogens and antigens is also reduced in human studies, and in animal models, such as the one of chronic protein-calorie deficiency developed by Martin.[11] Unlike some of the other abnormalities caused by malnutrition, the depression in lymphocyte count corrects rapidly with refeeding.

Respiratory tract IgA antibody

IgA is the major antibody of the lower respiratory tract and is produced locally in the lung. Children with protein calorie malnutrition have reduced levels of

secretory IgA in their saliva and nasopharyngeal secretions.[24] In adults with mal-
nutrition, we have observed that sputum levels of IgA were related to the PNI, with
IgA levels being reduced in patients with impaired nutritional status, and a high
PNI.[17] Complement levels may also be reduced with nutritional depletion (Martin).

Summary (Fig. 1)

Malnutrition has multiple effects on the host defense system, as outlined above.
When protein calorie deficiency is present, it can lead to an increased number of
upper and lower respiratory tract receptors for gram-negative bacteria. If this
enhanced mucosal susceptibility to bacterial binding occurs in a setting of prolonged bacterial contact with mucosal cells, as is the case when other airway host
defense fail, then colonization may result. When nutritional impairment is present
and leads to not only colonization, but also to a loss in lung phagocytic function,
humoral immune function, and delayed type hypersensitivity, then colonization
may proceed to pneumonia.

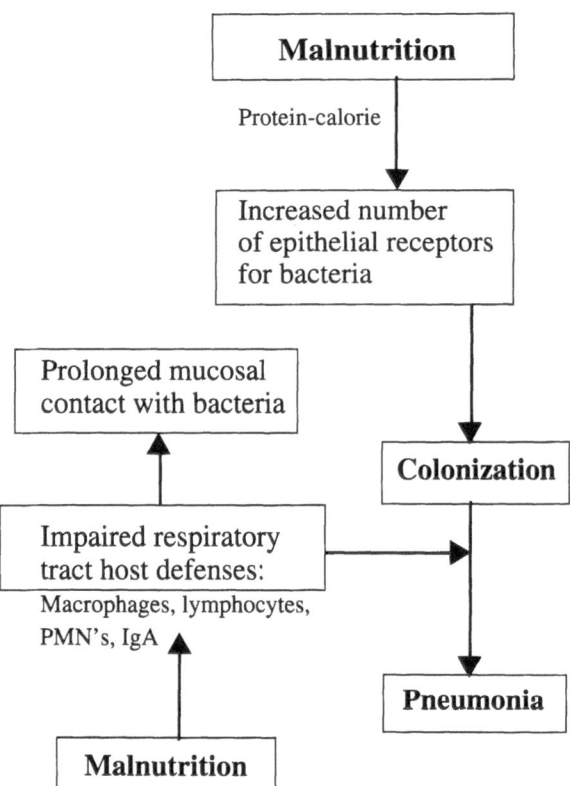

Fig. 1. The Malnutrition flow-chart.

Nutritional Therapy and Pneumonia

The intestinal tract as a source of pneumonic pathogens

The intestinal tract, and the stomach in particular, may be overgrown with gram-negative bacteria and when this occurs, the stomach may serve as a reservoir for organisms that can be transmitted to the lung and lead to pneumonia.[2,7,22] The extent to which the stomach is colonized by gram-negative bacteria depends on gastric pH. When pH rises, as may occur with the administration of antacids, histamine type 2 antagonists, or even enteral feeding, the number of gram-negative bacteria in the stomach increases.

At a pH of 2, the stomach may have no more than 100 gram-negative organisms per ml. of gastric juice, while at a pH of 6, the bacterial count can increase as high as 10^6 to 10^8 organisms per ml.[2] Mechanically ventilated patients frequently have gastric colonization, and in one study, gastric to tracheal transmission of bacteria occurred in 17 of 52 such individuals.[7]

Factors that promote this type of transmission include paralytic ileus, nasogastric tubes and gastric acid neutralization.[21]

Enteral feedings may also add to the risk of inoculating intestinal tract bacteria into the lung. Pingleton studied 16 patients treated with enteral nutrition, but without acid neutralizing therapy.[22] Gastric pH was high in this population[5,6], but did not increase with feeding. Initially 12 gram-negative isolates were recovered from these patients, but this number increased to 24 after 2 days of enteral feeding. Five patients had evidence for gastric to tracheal transmission of bacteria, and in four of these patients, gram-negative organisms were involved.

Several mechanisms may explain why enteral feeding can be complicated by gastric to tracheal transmission of bacteria.

The presence of a feeding tube may serve to breach the integrity of the gastro-oesophageal junction and provide a pathway for organisms to travel out of the stomach to the oropharynx, where they may then be aspirated around the endotracheal tube cuff.

Depending on the size of the feeding tube, the location of its tip, the type of endotracheal tube cuff used, and the method of feeding (intermittent bolus or continuous infusion), the chance of aspiration will vary widely. With low volume, high pressure endotracheal tube cuffs, aspiration may occur in 56% of all patients, but with high volume cuffs, this rate can be reduced to 20%.[21] Bolus feeding, even with small bore feeding tubes, can be associated with an aspiration rate as high as 38%, but when these tubes are used with continuous infusion feedings, the rate of aspiration is as low as 0-5.7%.[21]

Even if a small tube is used with a continuous feeding technique, care must be taken to recognize and avoid accidental dislodgement, which is common and may add to risk of aspiration.

Nutritional Therapy to Prevent Pneumonia

Many of the host defense deficits induced by malnutrition are reversible with refeeding, but the time required for this reversal may vary depending on the abnormality present. Bacterial adherence to airway epithelial cells can be reduced with feeding, and this may occur in as short a period as two days, in animal models of starvation, but may require weeks in malnourished hospitalized adults.[9,18] Delayed type hypersensitivity abnormalities appear to reverse rapidly, within days of refeeding, while macrophage and neutrophil deficits may require several weeks to disappear.[1,4,11,23] Even if the abnormalities can be ameliorated, it is still unclear if the reversals seen in animal models and non-infected adults, can occur in acutely ill patients with both malnutrition and pneumonia. If reversal cannot occur in this population, then nutritional therapy may not be useful as an adjunct to antimicrobial medications, in the presence of established infection.

Nutritional therapy may, however, be useful to prevent infection if it is administered to patients, at risk for protein-calorie depletion, prior to the advent of pneumonia. Although this prophylactic approach has not been widely applied, Moore et al. have used this type of nutritional therapy in one at risk population, those with major abdominal trauma.[12] In this study, 59 patients were evaluated and randomized to receive either enteral feeding or total parenteral nutrition. Since all patients had an exploratory laparotomy, the enteral feeding group received nutritional therapy through a feeding jejunostomy, thereby avoiding all of the risks of nutritional therapy into the stomach. In all, 29 patients received enteral feeding, with an intake of 1847 calories, while 30 received total parenteral nutrition, with an intake 2261 calories. Pneumonia occurred in 6 patients, and all of them were in the group receiving total parenteral nutrition. The Authors concluded that parenteral nutrition was an independent risk factor for pneumonia, while, possibly, enteral feeding was able to prevent infection. This prevention did not occur solely as a consequence of caloric replacement, because patients receiving adequate calories parenterally still developed pneumonia. Rather, by feeding patients enterally, the intestinal mucosa was stimulated and not allowed to atrophy. This in turn may have prevented bacterial translocation across the mucosa which can be a source of systemic sepsis and multi-system organ failure. In addition, with enteral feeding, the liver may make fewer inflammatory mediators than it does in states of injury when the intestine is not fed. Because these mediators may themselves cause multi-system organ failure, especially in the presence of endotoxemia following bacterial translocation, the influence of enteral feeding may be particularly beneficial.[12] Thus, the challenge of the future is not only to determine the deficits in host defense that follow malnutrition, but also to establish whether feeding can be used to minimize the clinical risk of pneumonia. As several recent studies have shown, not all refeeding programs can avoid infection, and attention must be given to the type of feeding given, as well as the methods used to deliver this feeding.

References

1. Anderson D.C., Krishna G.S., Hughes B.J., et al.: Impaired polymorphonuclear leukocyte motility in malnourished infants: relationship to functional abnormalities of cell adherence. J. Lab. Clin. Med. 1983; 101:881-895
2. Atherton S.T., White D.J.: Stomach as source of bacteria colonizing respiratory tract during artificial ventilation. Lancet 1978; 2: 968-969
3. Brown K.H., Gilman R.H., Gaffar A., et al.: Infections associated with severe protein-calorie malnutrition in hospitalized infants and children. Nutr. Res. 1981; 1:33-46
4. Chandra R.K.: Cell-mediated immunitary in nutritional imbalance. Federation. Proc. 1980; 39:3088-3092
5. Driver A.G., LeBrun M.: Iatrogenic malnutrition in patients receiving ventilatory support. JAMA 1980; 244:2195-2196
6. Driver A.G., McAlevy M.T., Smith J.L.: Nutritional assessment of patients with chronic obstructive pulmonary disease and acute respiratory failure. Chest 1982; 82:568-571
7. DuMoulin G.C., Hedly-Whyte J., Paterson D.G., et al.: Aspiration of gastric bacteria in antiacid-treated patients: A frequent cause of postoperative colonization of the airway. Lancet 1982; 1:242-245
8. Green G.M., Kass E.H.: Factors influencing the clearance of bacteria by the lung. J. Clin. Invest. 1964; 43:769-776
9. Higuchi J.H., Johanson W.G.: The relationship between adherence of *Pseudomonas aeruginosa* to upper respiratory cells *in vitro* and susceptibility to colonization *in vivo*. J. Lab. Clin. Med. 1980; 95:698-705
10. Martin T.R.: The relationship between malnutrition and lung infections. Clin. Chest. Med. 1987; 8:359-372
11. Martin T.R., Altman L.C., Alvares O.F.: The effects of severe protein-calorie malnutrition on antibacterial defense mechanisms in the rat lung. Am. Rev. Respir. Dis. 1983; 128:1013-1019
12. Moore F.A., Moore E.E., Jones T.N., McCroskey B.L., Peterson V.M.: TEN versus TPN following major abdominal trauma-reduced septic mortality. J. Trauma 1989; 29:916-923
13. Mullen J.L., Gertner M.H., Buzby G.P., Goodhart G.L., Rosato E.F.: Implications of malnutritional in the surgical patient. Arch. Surt. 1979; 114:121-125
14. Niederman M.S.: Bacterial adherence as a mechanism of airway colonization. Eur. J. Clin. Microbiol. Infect. 1989; 8:15-20
15. Niederman M.S., Fein A.M.: The interaction of infection and the adult respiratory distress syndrome. Crit. Care Clinics 1986; 2:471-495
16. Niederman M.S., Mantovani R., Schoch P., Papas J., Fein A.M.: Patterns and routes of tracheobronchial colonization in mechanically ventilated patients: the role of nutritional status in colonization of the lower airway by Pseudomonas species. Chest 1989; 95:155-161
17. Niederman M.S., Merrill W.W., Polomski S., Gee J.B.L., Reynolds H.Y.: Influence of sputum IgA and elastase on tracheal cell bacterial adherence. Am. Rev. Respir. Dis. 1986; 133:255-260
18. Niederman M.S., Merrill W.W., Ferranti R.D., Pagano K.M., Palmer L.B., Reynolds H.Y.: Nutritional status and bacterial binding in the lower respiratory tract in chronic tracheostomy. Ann. Inter. Med. 1984; 100:795-800
19. Nwiloh J., Freeman H., McCord C.: Malnutrition: an important determinant of fatal outcome in surgically treated pulmonary suppurative disease. J. Natl. Med. Assoc. 1989; 81:525-529

20. Phillips I., Wharton F.: Acute bacterial infection in kwashiorkor and marasmus. Br. Med. J. 1968; 407-409

21. Pingleton S.K.: Enteral nutrition as a risk factor for nosocomial pneumonia. Eur. J. Clin. Microbiol. Infect. Dis. 1989; 8:51-55

22. Pingleton S.K., Hinthorn D.R., Liu C.: Enteral nutrition in patients receiving mechanical ventilation: multiple sources of tracheal colonization include the stomach. Am. J. Med. 1986 80:827-832

23. Shennib H., Chiu R.C.J., Mulder D.S., Lough J.O.: Depression and delayed recovery of alveolar macrophage function during starvation and refeeding. Surg. Gynecol. Obstet. 1984; 158:535-540

24. Stiehm E.R.: Humoral immunity in malnutrition. Federation Proc. 1980; 39:3093-3097

10. Respiratory Muscle Function in the Obese

M.G. SAMPSON,[1] A.GRASSINO[2]

1. Department of Medicine, Division of Pulmonary Disease, State University of New York, Stony Brook, New York, USA
2. Meakins-Christie Laboratory, Royal Victoria Hospital, McGill University and Notre Dame Hospital, University of Montreal, Quebec, Canada.

Introduction

Obese patients perform poorly at task requiring increased ventilatory levels such as during maximum voluntary ventilation (MVV) manoeuvres and exercise.[1,2] Their increased chest wall load with its attendant reduction in chest wall compliance causes increased work of breathing at all levels of ventilation. In addition, it has been reported that obese patients have reduced inspiratory muscle strength as assessed by peak inspiratory mouth pressures and transdiaphragmatic pressures.[3,4] Normal subjects have been shown[5,6] capable of sustaining minute ventilation levels ranging from 55 to 80% of their predicted maximum voluntary ventilation (MVV). More recent data[7] suggest that normal subjects will demonstrate electromyographic and mechanical diaphragmatic fatigue at ventilatory levels above 76% MVV with diaphragmatic tension-time indices (TtDi) greater than 0.15. Given the reduced chest wall compliance and inspiratory muscle strength typical of the obese, we hypothesized that their poor mechanical ventilatory coupling should cause fatigue of their diaphragms at levels of ventilation well below that reported in normals, i.e., at levels of less than 76% of MVV.

Methods

Fourteen obese eucapnic male patients ranging in weight from 130% to 275% ideal body weight (%IDBW) and 5 normal weight males were studied in the seated position during progressive hyperoxic hypercapnic hyperventilation.[8] The obese patients were divided into two subgroups based on their %IDBW as follows: Group

1A (n = 6) were morbidly obese, i.e., %IDBW > 200%; Group 1B (n = 8) were moderately obese with %IDBW from 130% to 200%. %IDBW was calculated based on height and sex from predicted normal weights.[9]

All patients were eucapnic at rest.

The patients/subjects were seated while on a mouth piece, with the following parameters measured:

1. flow at the mouth from which tidal volume, breathing frequency and minute ventilation were obtained;
2. pressure at the mouth;
3. pleural (Ppl) and gastric pressures (Pga) via latex balloon-catheter systems[10] from which transdiaphragmatic pressure (Pdi) was measured by a differential pressure transducer (pdi = Pga - Ppl);
4. electromyogram of the diaphragm (EMGdi) via an oesophageal bipolar electrode.[11]

End tidal CO_2 was measured continuously. P 0.1, defined as the pressure developed at the mouth 100 msec after the onset of an occluded inspiration[12], was measured via a silent valve. Maximum transdiaphragmatic pressure (PdiMAX) was obtained by having the patient/subject inspire maximally against a closed airway at FRC.[13] The tension -time index per breath for the diaphragm (TTdi) times the duty cycle, or, TtDi = % Pdi * (Ti/Ttot).[14,15] The diaphragmatic EMG was processed breath by breath for total, rectified, integrated activity (Edi) via a leaky integrator circuit.[16] Edi was expressed as a percentage of the maximum Edi obtained during a TLC manoeuvre, %Edi. The EMGdi was also analysed for its high (H) and low (L) frequency components[16]; the EMGdi H/L ratio was normalized for each run by expressing it in percent of the maximum H/L attained during the run, % H/L.

Results

Group 1A patients had reduced PdiMax, predicted MVV and PaO_2 in comparison with the normal group. The three groups demonstrated similar P. 100/$PetCO_2$ responses; and, as expected, due to their mechanical disadvantage, obese patients had blunted ventilatory (%MVV/$PetCO_2$) responses. Neural-ventilatory coupling (%MVV vs P. 100) was similar in the two obese groups but significantly lower than in the normals (Table I).

As shown on figure 1, the obese had reduced % MVV during the later part of their runs in comparison with the normal subjects. There were no differences in the Ti/Ttot responses which remained relatively fixed at values of 0.42 to 0.46 in all patients groups throughout the hyperpnoea runs. 4 of 6 Group 1A patients had their %MVV fall during the last (sixth) minute of their hyperpnoea runs. The moderately obese Group 1B demonstrated no "ventilatory" failure at any point during their

Table I Mean value (+/- 1 S.D.) for three groups (defined in Methods).

Parameter % IDBW		Group 1A > 200 %		Group 1B > 130 %		Normals < 130 %
	(1A vs Norm)		(1A vs 1B)		(1B vs Norm)	
	|		|		|	
	V		V		V	
PdiMAX	*	78 (17)	*	105 (46)	*	153 (23)
(P.100) / (PetCO$_2$)	ns	.63 (.36)	ns	0.77 (.44)	ns	0.65 (.06)
P.100 @ 55mmHg	ns	7.0 (3.0	ns	6.6 (3.2)	ns	4.2 (1.2)
(% MVV) / (PetCO$_2$)	*	1.1 (.12)	ns	1.7 (1.0)	*	3.0 (1.4)
(% MVV) / (P;100)	*	1.7 (1.0)	ns	2.2 (1.1)	*	4.7 (2.1)
TF (secs)	*	368 (99)	*	627 (160)	*	> 1,000

Parameter abbreviations are defined as follows: PdiMAX = Maximum Transdiaphragmatic Pressure; P.100 = Mouth Occlusion Pressure at 100 msec; % MVV = Percent of predicted maximum voluntary ventilation; PetCO$_2$ = End-tidal CO$_2$; TF = Diaphragmatic EMG Fatigue Constant. Comparison by ANOVA using Tukey tecnique with * = significant at P < .05 level; ns = not significant.

hyperpnoea runs. The mean P. 100 at any given minute during the hyperpnoea runs was similar in all three groups.

The average %Pdi responses during the hyperpnoea runs for the 2 obese groups, in comparison with the normal subjects, were significantly greater throughout the majority of the hyperpnoea runs (Fig. 1b). The %Edi response for the morbidly obese was similar to that of the normals. The ratio of %Pdi to %Edi, a reflection of diaphragmatic neural-mechanical coupling efficiency, was larger in the obese than in normals (Fig. 1c). During the last minute of their hyperpnoea runs, 2 of 6 Group 1A patients and 3 of the 8 group 1B patients demonstrated a fall in this ratio; i.e., these patients developed neural-mechanical coupling failure consistent with mechanical diaphragmatic fatigue.

The average TtDi responses during the hyperpnoea runs was larger in the obese groups (Fig. 2a). The obese groups demonstrated marked and significant falls in their EMGdi H/L ratio during the hyperpnoea runs whereas the normal subjects did not (Fig. 2b). In cases where EMGdi H/L fell to less than 85% control, an exponential function was fitted to the decaying phase of each run. The rate of decay of the H/L was then expressed as the time constant derived from the exponential equation for each run.[15,17] This value is referred to as the fatigue time constant (TF) and its average value for the 2 obese groups is shown in Table I. The morbidly obese group of obese patients had the shortest TF and the highest TtDi in comparison with

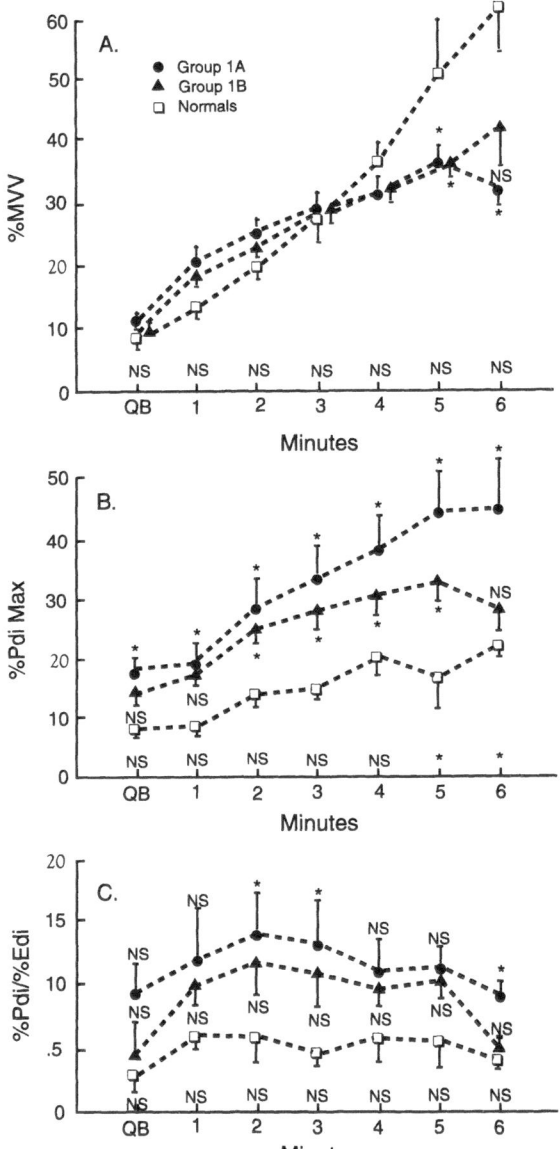

Fig. 1 Averaged values for three groups (Defined in Methods) at each minute during hyperpnea runs. Bars = ±1SE. P values shown immediately above or below Group 1A/1B SE bars represent comparison between normal group and group 1A/1B, respectively. P value shown above x-axis is for comparison between group 1A and 1B; comparison by ANOVA using Tukey technique with * = significant at P .05 level; ns = not significant; if not shown = ns

Panel a: Average % maximum voluntary ventilation (%MVV);

Panel b: Average transdiaphragmatic pressure response;

Panel c: Average %Pdi to %Edi ratio (%Pdi/%Edi).

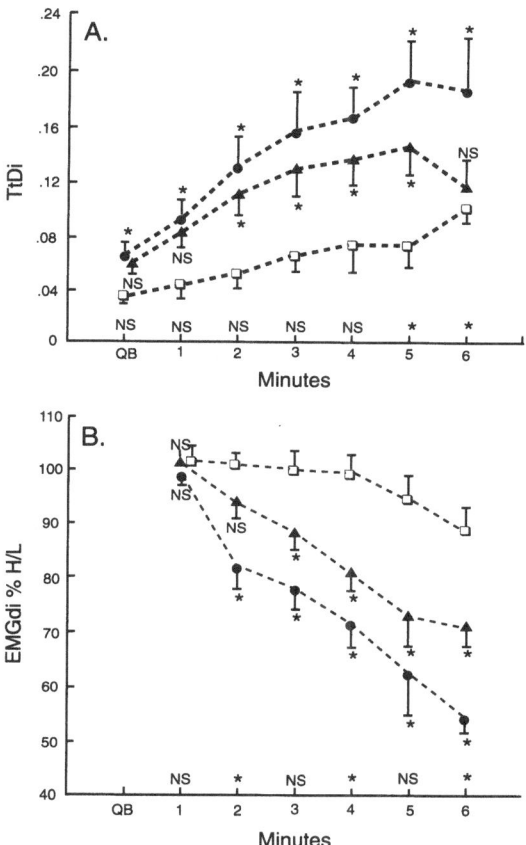

Fig. 2 Averaged values for three groups (Defined in Methods). Bars and P values are as described for Figure 1.
Panel a: Average DiaphragmaticTension-Time Index (TtDi);
Panel b: Average ratio of high to low frequency power (H/L ratio) of diaphragmatic electromyogram (EMGdi).

the lesser obese group and the normal subjects.

The advent of electrical fatigue of the diaphragm (defined as a fall in EMGdi H/L to below 80% control) correlated with the TtDi level (Fig. 3a). Diaphragmatic Tension-time indices of less than 0.10 were not associated with electrical diaphragmatic fatigue. As the level of hyperpnoea increased (Fig. 3b), every obese patient had their TtDi cross over their diaphragmatic Tension-time fatigue threshold and electrical fatigue occurred. The extent of fall in their EMGdi H/L ratio correlated with the TtDi level with those patients developing the largest TtDi indices demonstrating the greatest and quickest falls in their EMGdi H/L ratios (Fig. 3c). The level of minute ventilation at which EMGdi fatigue occurred, varied between the three groups: normal subjects while generating 60% MVV demon-

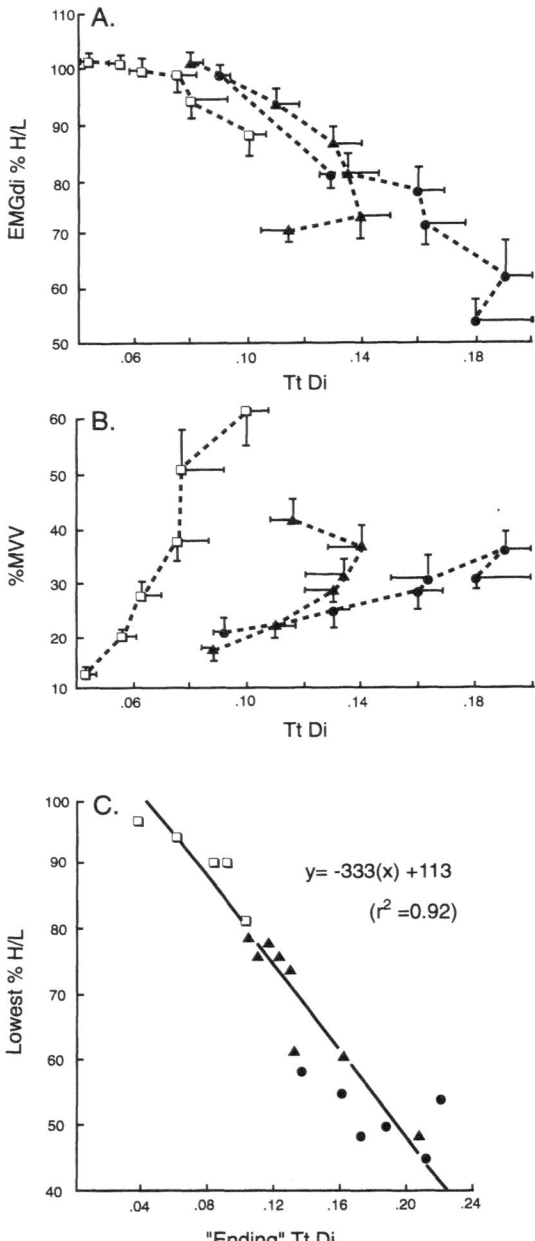

Fig. 3 Panels a and b: Averaged values for % H/L and % MVV for three groups (defined above) at progressively larger TtDis. Bars and abbreviations as in Figure 1 and 2.

Panel c: Lowest % H/L obtained during hyperpnoea run is plotted versus the average "ending" TtDi generated during the last 3 minutes. Each point represents an individual's run. Symbols as previously described. Linear regression analysis shown on figure.

strated no evidence of fatigue whereas the more obese patients at rather trivial levels of 27% MVV (Group 1A) and 37% MVV (Group 1B) demonstrated electrical fatigue of their diaphragms and in five cases mechanical failure was also apparent.

Discussion

The reduction in ventilatory response during carbon dioxide rebreathing (Fig. 1A) was not a result of blunted "central" neural drive as both obese groups demonstrated normal p. 100 and %Edi responses and it probably relates to chest wall mechanical factors.[8,18,19] The important mechanical factors affecting the ventilatory capability in the obese are the load imposed on their chest walls and intrinsic respiratory muscle properties.[1,3] The above observations are equally applicable to other respiratory disorders which cause increased mechanical loads, e.g., chronic obstructive pulmonary disease with increased resistive airway loads[20] and interstitial pulmonary fibrosis patients with increased lung parenchymal loads.[21,22]

In the present study, eucapnic obese patients were found to have better than normal neuro-mechanical coupling (i.e. % Pdi/%Edi ratio) which we believe was a result of their improved diaphragmatic operating lengths (lower %FRC). However, these patients also demonstrated significantly reduced PdiMAX (Table I).

In the absence of neuromuscular disease it is not clear how this diaphragmatic muscular weakness develops in the obese but it has been observed by others.[1,19] This intrinsic diaphragmatic muscle reserve deficiency coupled with the increased chest wall loads found in the obese places them at increased risk of diaphragmatic failure.

Obese patients were found to have substantially greater %Pdi and TtDi in comparison with the normals (Figs. 1b, 2a). Obese patients were generating greater than normal diaphragmatic force to achieve any given level of %MVV (Fig. 3b). Although the obese patients demonstrated better than normal diaphragmatic neuro-mechanical coupling (Fig. 1c), they all developed evidence of diaphragmatic electrical fatigue by EMGdi H/L criteria. Bellemare and Grassino[15,17] have shown that there is a critical TtDi of 0.15 to 0.18 above which fatigue of the diaphragm will occur. Our obese patients developed evidence of electrical diaphragmatic fatigue, i.e., falls in their EMGdi H/L ratio to less than 80% control, at an average TtDi of 0.14 (range = 0.11 to 0.16) (Fig. 3a). Our normal subjects did not achieve TtDi of greater than 0.11 at any time and did not fatigue. The critical fatiguing TtDi from our study would therefore be placed in the range of 0.11 to 0.16. We also demonstrated a relationship between TtDi and the diaphragmatic decay constant (TF) remarkably similar to that observed by Bellemare et al.[15] (Fig. 3c)

Conclusion

We have demonstrated that obese patients are more susceptible to diaphragmatic muscle fatigue than normal subjects during tasks requiring levels of minute ventilation as low as 40% MVV. A fall in the H/L ratio of the diaphragmatic EMG in all the obese patients and worsening diaphragmatic neural-mechanical coupling in 5 of the obese provide evidence that fatigue of the diaphragm was occurring in these patients. Despite markedly larger minute ventilation levels of over 60% MVV, normal subjects, as shown by others[5,7], did not demonstrate diaphragmatic fatigue; thus, the advent of diaphragmatic fatigue was not uniquely related to the level of hyperpnoea (Fig. 3b). In fact, we observed that falls in diaphragmatic EMG H/L ratio and the diaphragmatic fatigue time constant (TF) were strongly related to the diaphragmatic TtDi (Fig. 3c). Our observations suggest that obese patients develop electrical fatigue of their diaphragms at a critical TtDi in the range of 0.11 to 0.16 at % MVV of less than 40% maximum.

References

1. Rochester D.F., Enson Y.: Current concepts in the pathogenesis of the obesity-hypoventilation syndrome. Am. J. Med. 1974; 57:402-20
2. Farebrother M.J.B.: Respiratory function and cardiorespiratory response to exercise in obesity. Br. J. Dis. Chest 1979; 73:211-29
3. Sharp J.T., Sweany S.K., Meadows W.R., Pietras R.J.: The Total Work of breathing in normal and obese men. J. Clin. Invest. 1964; 43:728-39
4. Sharp J.T., Druz W.S., Kondragunta V.R.: Diaphragmatic responses to body position changes in obese patients with obstructive sleep apnea. Am. Rev. Resp. Dis. 1986; 133:32-37
5. Belman M.J., Mirsky M., Hary D.: Diaphragmatic Fatigue during Sustained Hyperpnea. Am. Rev. Dis. 1981; 123:210-19
6. Freedman S.: Sustained maximum voluntary ventilation. Respir. Physiol. 1970; 8:230-44
7. Bai T.R., Rabinovitch B.J., Pardy R.L.: Near-maximal voluntary hyperpnea and ventilatory muscle function. J. Appl. Physiol. 1984; 57:1742-48
8. Sampson M.G., Grassino A.E.: Neurochemical properties in obese patients during carbon dioxide rebreathing. Am. J. Med. 1983; 75:81-90
9. Metropolitan Life Insurance Company: New weights and standards for men and women. Build study, Society of Actuaries and Association of life insurance medical directors of America, 1979
10. Milic-Emili J., Mead J., Turner J.M., Glausser E.M.: Improved Technique for estimating pressure from esophageal balloons. J. Appl. Physiol. 1964; 19:207-11
11. Grassino A.E., Whitelaw W.A., Milic-Emili J.: Influence of lung volume and electrode position on electromyography of the diaphragm. J. Appl. Physiol. 1976; 40:971-75
12. Whitelaw W.A., Derenne J.P., Milic-Emili J.: Occlusion pressure as a measure of respiratory center output in conscious man. Respir. Physiol. 1975; 23:181-99
13. Laporta D., Grassino A.: Assessment of Transdiaphragmatic pressure in humans. J. Appl. Physiol. 1985; 58:1469-76
14. Bellemare F., Grassino A.: Evaluation of human diaphragm fatigue. J. Appl. Physiol. 1982; 53:1196-1206
15. Bellemare F., Grassino A.: Effects of pressure and timing of contraction on human diaphragm fatigue. J. Appl. Physiol. 1982; 53:1190-95
16. Gross D., Grassino A., Ross W.R.D., Macklem P.T.: Electromyographic pattern of diaphragmatic fatigue. J. Appl. Physiol. 1979; 46:11-7

17. Bellemare F., Grassino A.: Force reserve of the diaphragm in patients with chronic obstructive pulmonary disease. J. Appl. Physiol. 1983; 55:8-15

18. Lopata M., Onal E.: Mass loading, sleep apnea, and the pathogenesis of obesity-hypoventilation. Am. Rev. Resp. Dis. 1982; 126:640-45

19. Sampson M.G., Grassino A.E.: Load compensation in obese patients during quiet tidal breathing. J. Appl. Physiol. 1983; 55:1269-76

20. Aubier M., Murciano D., Fournier M., Milic-Emili J., Pariente R., Derenne J-P. Central respiratory drive in acute respiratory failure of patients with chronic obstructive pulmonary disease. Am. Rev. Resp. Dis. 1980; 122:191-99

21. DiMarco A.F., Kelsen S.G., Cherniack N.S., Gothe B.: Occlusion pressure and breathing pattern in patients with interstitial lung disease. Am. Rev. Resp. Dis. 1983; 127:425-430

22. Renzi G., Milic-Emili J., Grassino A.E.: The pattern of breathing in diffuse lung fibrosis. Bull. Eur. Physiopathol. Respir. 1982; 18:461-72

11. Eating Behavior in Chronic Obstructive Pulmonary Disease

G. Maiani,[1] S. Callegari,[1] L. Viola,[1] A. Calvi,[2] T. Montagna[3]

Clinica del Lavoro Foundation, Institute of Care and Research Medical Center of Rehabilitation, Montescano, Pavia, Italy
1 Psychology Unit
2 Nutritional Unit
3 Pneumology Department

The psychological aspects of nutrition problems in COPD patients are almost never considered by literature. What is more, eating behavior has hardly ever been taken into account in evaluating these patients' Life Quality.

A first example of this can be found in the Chronic Respiratory Disease Questionnaire by Guyatt et al.[3]

It includes items of dyspnoea, mastery (the degree to which a patients feels control over the disease or its manifestations), fatigue, sleep disturbance, emotional disorders, social difficulties and cognitive problems. Eating is not specifically considered. It does appear, but only among the activities that might cause dyspnoea.

A second instrument, the Bronchitis Emphysema Symptom Checklist[4] allows us to understand how patients cope with and experience chronic bronchitis or emphysema. It consists of 11 symptom categories and more than fifty items, but only two of them refer to eating behavior. These are included in the Decathexis cluster, which measures the patient's lack of interest towards the routine aspects of life: food, things and other people. A specific assessment of eating behavior is however omitted.

Finally, a third instrument, the Sickness Impact Profile[2] allows the assessment of twelve behavioral areas, grouped into three main categories: Physical, Psycho-social and Independent Categories. Here eating behavior is specifically assessed.

A study on 166 severe COPD patients compared to normal subjects by means of the Sickness Impact Profile,[5] shows that the life quality of COPD patients is generally impaired, but that eating seems to be the only area in which COPD patients have no difficulties if compared to normals. We must however consider that these data refer to the whole sample and also, that authors say nothing whatso ever about nutritional status in these patients.

In order to understand if nutritional status is in some way linked to psychological variables in COPD we studied 62 patients consecutively admitted to our Psychology Unit for routine assessment. Patients were classified as COPD according to the American Thoracic Society standard criteria[1] and had functional signs of bronchial obstruction, with an FEV_1 less than 80% predicted.

They were divided into 4 groups according to their Body Mass Index as proposed by the Italian Institute of Nutrition.[6]

By means of this method we found that 12.9% were underweight, 40.3% normal weight, 29% overweight and 17.7% obese.

Their age and educational level are shown in table I. Obese patients resulted in being younger than underweight patients.

The majority were males and had a low or medium education. All patients completed Primary Scales of the Cognitive Behavioral Assessment[7] that provides scores on several psychologic and behavioral variables (anxiety, personality characteristics, psychophysiological disorders, fear and phobias, depression, obsessive and compulsive traits).

The patients also described their eating behavior by choosing one or more of the given statements and finally, were asked to say if they were bothered by one or more

Table I. Sample characteristics.

| | Age | Male | Female | Educational level | | |
				low	medium	high
Under-weight n = 8	64.75±4.02	6 (75.0%)	2 (25.0%)	5 (62.5%)	2 (25.0%)	1 (12.5%) .
Normal weight n = 25	58.84±6.99	20 (80.0%)	5 (20.0%)	16 (64.0%)	7 (28.0%)	2 (8.0%)
Over-weight n = 18	60.44±6.63	13 (72.2%)	5 (27.7%)	9 (50.0%)	7 (38.9%)	2 (11.1%)
Obese n = 11	54.73±9.55	7 (63.6%)	4 (36.7%)	8 (72.7%)	3 (27.3%)	0 -

F: 3.25 p=.02 (obese/u.w.) (d.f.: 3;58)

of the given gastric and respiratory symptoms.

Our results suggest that there is no relationship between basic psychological variables and body weight: the Anova test on Cognitive Behavioral Scales obtained from the four groups showed no significant differences (Table II).

As to the description of their eating behavior (Table III) patients gave responses that can be summarized as follows:

Table II. Cognitive Behavioral Assessment (Primary Scales): mean and standard deviation

C.B.A. Scales	Under-weight	Normal weight	Over-weight	Obese	F
	Mean (SD)	Mean (SD)	Mean (SD)	Mean (SD)	
STAI X-1 State anxiety	.21 (1.02)	.69 (1.13)	.24 (0.83)	.10 (0.56)	1.28
STAI X-2 Trait anxiety	.46 (1.23)	.47 (0.86)	.15 (1.07)	.15 (0.75)	0.53
EPQ E Extraversion	.54 (1.02)	.23 (0.99)	.48 (0.91)	.63 (0.79)	0.55
EPQ N Emotional Instability	-.15 (1.13)	-.23 (1.08)	-.30 (0.94)	-.27 (1.19)	0.03
EPQ P Antisocial Traits	1.42 (1.73)	1.20 (1.19)	1.44 (1.61)	1.78 (1.84)	0.35
EPQ L Lie	.76 (0.50)	.10 (1.06)	.28 (0.83)	.47 (0.46)	1.32
QPF Psychophysiological disorders	.91 (1.32)	.72 (1.40)	.17 (0.98)	.08 (0.65)	1.47
IP F Fear and phobias overall	.03 (0.61)	.07 (0.84)	.04 (0.94)	.11 (0.89)	0.02

IP 1 Fear of calamities	.14 (0.87)	-.12 (0.81)	.07 (1.06)	.11 (0.83)	0.28
IP 2 Social fears	.07 (0.99)	.05 (0.98)	.05 (1.01)	.08 (0.72)	0.00
IP 3 Fear of repellent animals	-.31 (0.62)	-.02 (0.85)	-.12 (1.04)	-.04 (1.23)	0.18
IP4 Agoraphobic traits	.24 (1.37)	.73 (1.54)	.38 (1.32)	.21 (1.16)	0.48
IP5 Medical fears	.08 (0.64)	.003 (0.85)	-.30 (0.57)	.34 (1.19)	1.31
QD Depression	.44 (0.81)	.29 (0.89)	.05 (0.88)	.15 (0.57)	0.50
MOCQ R obsessive/compulsive traits (overall)	.14 (1.02)	.38 (0.13)	.28 (1.23)	.34 (1.26)	0.09
MOCQ 1 Checking	-.22 (1.03)	.11 (1.17)	.29 (1.17)	.33 (1.11)	0.44
MOCQ 2 Cleaning	.68 (1.09)	.78 (1.24)	.25 (1.05)	.09 (1.39)	1.13
MOCQ 3 Doubting ruminating	-.45 (1.17)	-.34 (0.84)	-.26 (0.93)	.11 (0.78)	0.74

P = not significant NS

- half of the *Underweight* patients say they have little appetite, whereas only one of them says he eats less when tired or nervous;
- *normal Weight* patients almost never refer to a lack or excess of appetite;
- some *Overweight* patients say they eat too fast and ought to lose weight. A few of them say eat too much;
- many *Obese* patients say they eat too fast and ought to be on a diet, but only a few of them say they eat too much, and none when tired or nervous.

Finally, as shown in Table IV, underweight patients complain more of heaviness, abdominal pain and nausea than the other patients. Furthermore, they refer more respiratory difficulties, even if this is not objectively confirmed by our lab tests.

To conclude, we would like to point out a couple of considerations.

First of all, eating disorders in response to emotional stimuli may affect the nutritional status of underweight patients in which it is already impaired by both hypermetabolism and gastric symptoms. This is the reason why the patient's response style to emotional states must be diagnosed as soon as possible.

Secondly, obese patients show a general lack of control in eating behavior. Furthermore they may not recognise or deny any connection between emotional stimuli and food intake.

Health education together with eating behavior functional analysis is therefore of utmost importance and we can safely say that eating behavior does not often seem disturbed by psychological factors in COPD patients. Nevertheless when it does in fact happen the situation can get very complicated. Therefore this is why in our

Table III. Describe your eating habits by selecting one or more of these statements.

Statements	answers: n (%)			
	U.W.	N.W.	O.W.	OB.
1. I usually have breakfast and main meals	6 (75.0)	17 (68.0)	9 (50.0)	2 (18.2)
2. I hardly eat when I'm tired or nervous because my stomach tightens up	1 (12.5)	2 (8.0)	0 -	0 -
3. I eat very little, I'm never hungry	4 (50.0)	4 (16.0)	1 (5.5)	0 -
4. I continuously eat when I'm tired or nervous	0 -	1 (4.0)	1 (5.5)	0 -
5. I eat too much, I'm always very hungry	0 -	2 (8.0)	4 (22.2)	3 (27.3)
6. I often thought about eating less to lose weight	0 -	2 (8.0)	7 (38.9)	7 (63.6)
7. I tried to follow a diet in vain	0 -	1 (4.0)	2 (11.1)	3 (27.3)

8. I often eat between meals	0	1	1	4
	-	(4.0)	(5.5)	(36.4)
9. I eat quickly, and I always finish before the others	1 (12.5)	11 (44.0)	6 (33.3)	6 (54.5)

U.W.: underweight; N.W.: normal weight; O.W.: overweight; OB.: obese

Table IV. Which of these somatic symptoms apply to you ?

| | answers: n (%) | | | |
Symptoms	U.W.	N.W.	O.W.	OB.
1. A feeling of heaviness and tightness of the stomach	2 (25.0)	4 (16.0)	1 (5.5)	2 (18.2)
2. Burning and acidity of the stomach	2 (25.0)	7 (28.0)	2 (11.1)	2 (18.2)
3. Swollen	2 (25.0)	4 (16.0)	6 (33.3)	6 (54.5)
4. Abdominal pain	1 (12.5)	2 (8.0)	1 (5.5)	0 -
5. Nausea	2 (25.0)	4 (16.0)	1 (5.5)	0 -
6. Vomit	0 -	2 (8.0)	0 -	1 (9.1)
7. A hole in the stomach	1 (12.5)	0 -	2 (11.1)	0 -
8. Difficult breathing	7 (87.5)	19 (76.0)	13 (72.2)	7 (63.6)

U.W.: underweight; N.W.: normal weight; O.W.: overweight; OB.: obese opinion eating behavior should always be assessed as soon as possible in underweight, overweight and obese patients.

References

1. A.T.S.: Standards for the diagnosis and care of patients with chronic obstructive pulmonary disease (COPD) and Asthma. Am. Rev. Resp. Dis. 1987; 136 1:225-243
2. Bergner M., Robbit R.A., Carter N.B., Gilson B.S.: The Sickness Impact Profile: Development and Final Revision of Health Status Measure. Medical Care 1981; 19, 8:785-805
3. Guyatt G.H., Berman L.B., Townsend M., Pugsley S.O., Chambers L.W.: A measure of quality of life for clinical trials in chronic lung disease. Thorax 1987; 42:773-778
4. Kinsman R.A., Fernandez E., Schocket M., Dirks J.F., Covino N.A.: Multidimensional Analysis of the Symptoms of Chronic Bronchitis and Emphysema. J. Behav. Med. 1983; 6,4:339-357
5. McSweeny A.J., Heaton R.K., Grant I., Cugell D., Sooliday N., Timms R.: The Chronic Obstructive Pulmonary Disease; Socioemotional Adjustment and Life Quality. Chest 1980; 77, 2 (Suppl.):309-311
6. Noé D.: *L'antropometria nella valutazione dello stato nutrizionale. Schede informative.* Istituto Scotti Bassani. 1986; n. 3
7. Sanavio E., Bertolotti G., Michielin P., Vidotto G., Zotti A.M.: C.B.A. *2.0 Scale Primarie.* Manuale. Firenze, Organizzazioni Speciali, 1986

Clinical Aspects of Malnutrition in COPD

12. The Effects of Nutritional Support to Malnourished ambulatory Chronic Obstructive Pulmonary Disease Patients

O. Nørregaard

Department of Respiratory Medicine, Aarhus University Hospital, Denmark

Malnutrition is prevalent in patients suffering from advanced COPD,[1,2] and has been shown to be correlated to impaired pulmonary function[3,4,5] as well as to a decrease in life expectancy.[6]

Several theories have been proposed to explain the association of malnutrition and COPD, and not surprising a low caloric intake has been suggested as a cause of the malnourished state. A number of studies have, however, documented that in general malnourished COPD patients do not eat less than the recommended daily allowances. Actually, they tend to have an elevated caloric intake[7,8] in harmony with the reported elevated resting energy expenditure and the up to ten fold increased respiratory work load of these patients.

The malnourished state could thus be the result of a relatively insufficient energy intake, although it may be well above recommended daily allowances.

The idea that this imbalance could be reversed by nutritional support appeared to be a promising possibility.

However, this field of research is rather new, and the number of controlled studies focusing on the effects of nutritional support to ambulatory malnourished COPD patients is so far rather limited.[9-13]

Non-controlled studies[14,15] have reported impressive effects from a moderate increase in nutritional intake,. The data from the few controlled studies have in general not been able to confirm these findings, and have on several important questions pointed towards conflicting conclusions.

Most of the studies[9,11,12] report of an elevated resting energy expenditure in the magnitude of 2000 calories per day in malnourished patients with a body weight of around 80% of ideal body weight, corresponding to roughly 150-160 % of resting energy expenditure. In conflict with this Efthimiou et al. found[10] a caloric intake of only 1400 calories per day, almost 1000 calories less than what was found in the study of Otte et al.[13] in a group of comparable patients. It is not obvious why the patients in Efthimiou's study in contrast with the general pattern of these patients did not have an elevated energy intake.

The ability to increase food intake during periods of nutritional supplementation lasting from 6 to 13 weeks varied considerably. In three studies[9,12,13] the patients were not able to make a net increase of more than roughly 10 % as any further increase in the quantity of ingested supplement was followed by a corresponding reduction in the patients ordinary food. In the study of Otte[13] the patients managed to increase their food intake in the order of 400 calories of supplement, resulting in a total caloric intake of ca. 2700 calories, corresponding to 210% of Resting Energy Expenditure (REE). There is no revealed information about the composition, the flavour, the colour nor any other quality of the diet nor of differences in the characteristics of the patients that could explain these marked differences.

In general the resulting increase in body weight varied between 1 and 2 kgs, except for the increase of 4.2 kgs in the study of Efthimiou.[10]

The weight increase in the other published long term study (13 weeks)[11] was 1.5 kg. Only one study reported an increase in lean body mass.[12] Body weight returned to baseline values within four weeks after discontinuation of nutritional supplement in one study[13] in contrast with a remaining significant weight gain of 2 kgs, 3 months after the patients returned to a diet very similar to their normal diet.[10]

The effects of nutritional supplements on anthropometrics showed a diverging pattern. Two studies found no significant effects[9,13] two[10,12] reported a significant increase in triceps skin fold thickness, one[10] in arm muscle circumference and two in the sum of four skin folds.[11,12]

Albumin increased significantly in two studies,[10,12] while no other study observed any significant change in albumin, transferrin, serum electrolytes, blood lymphocyte count, T-helper/T-suppresser ratio, pha-stimulation nor total white blood cell count.

Two studies found no effects on the respiratory muscles.[9,12] Efthimiou[10] found a significant increase in PI_{max} and in PE_{max} during nutritional supplementation as well as in handgrip strength and in sternomastoid muscle contractility and fatiguability. Knowles[12] observed in a cross-over study an increase in PI_{max} and in pressure time product in the control group, and an increase during supplementation in one of the two groups, however, resulting in values not different from the control group values.

The results concerning pulmonary function are in agreement with respect to some of the parameters. Thus, none of the studies has found any significant increase in FEV_1, FVC nor in arterial blood gasses. MVV was found to decrease in two studies[9,11] and to increase in one.[12]

Well being, walking distance and the sensation of dyspnea deteriorated in one study[11] and improved in another.[10]

Conclusion

Most of the controlled studies observe that the malnourished COPD patients have an elevated energy intake during baseline conditions, and that it is possible to a smaller extent to increase that. This results in a significant increase in body weight, usually around 1 to 2 kgs, in some cases followed by an increase in serum albumin but not in any other biochemical nor immunologic parameter. The data on respiratory muscle strength and endurance, sensation of dyspnea and on well being during nutritional supplementation are conflicting, as some report significant improvements and others not.

At present it thus seems that nutritional support to malnourished COPD patients under certain well defined circumstances to certain patients can be beneficial.

Long term randomized studies are needed to define more clearly this group of patients, the optimal regimen and not least to evaluate the results with respect to quality of life and life expectancy.

References

1. Filley G., Bechwitt H., Reeves J., Mitchell R.: Chronic obstructive pulmonary disease. Am. J. Med. 1968; 44:26-38
2. Hunter A.M., Carry M.A., Larsh H.W.: The nutritional status of patients with chronic obstructive pulmonary disease. Am. Rev. Respir. Dis. 1981; 124:376-81
3. Burrows B.A., Niden A.H., Barclay W.R., Kasik J.E.: Chronic obstructive pulmonary disease: relationship of clinical and physical findings to the severity of airways obstruction. Am. Rev. Respir. Dis. 1964; 91:665-78
4. Renzetti A.D., McClement J.H., Litt B.D.: The Veterans Administration cooperative study of pulmonary function. Mortality in relation to respiratory function in chronic obstructive pulmonary disease. Am. J. Med. 1966; 41:115-29
5. Driver A.G., McAlvey M.T., Smith J.L.: Nutritional assessment of patients with chronic obstructive pulmonary disease and acute respiratory failure. Chest 1982; 82:568-71
6. Vandenbergh E., Van de Woestijne K.P., Gyselen A.: Weight changes in the terminal stages of chronic obstructive pulmonary disease. Am. Rev. Respir. Dis. 1967; 95:556-66
7. Openbrier D., Irwin M., Rogers R.M., et al.: Nutritional status and lung function in patients with emphysema and chronic bronchitis. Chest 1983; 83:17-22
8. Braun S.R., Dixon R.M., Keim N.L., Luby M., Anderegg A., Shrago E.S.: Predictive clinical value of nutritional asessment factors in COPD. Chest 1984; 85:353-7
9. Lewis M.I., Belman M.J., Dorr-Uyemura L.: Nutritional Supplementation in Ambulatory Patients with Chronic Obstructive Pulmonary Disease. Am. Rev. Respir. Dis. 1987; 135:1062-68

10. Efthimiou J., Fleming J., Gomes C., Spiro S.G.: The effect of supplementary oral nutrition in poorly nourished patients with chronic obstructive pulmonary disease. Am. Rev. Respir. Dis. 1988; 137:1075-82
11. Nørregaard O., Tottrup A., Saaek A., Hessov I.: Effects of oral nutritional supplements to adults with chronic obstructive pulmonary disease. Bull. Eur. Physiopathol. Respir. 1987; 23 (suppl. 12) 388
12. Knowles J.B., Fairbarn M.S., Wiggs B.J., Chan-Yan C., Pardy R.L.: Dietary supplementation and respiratory muscle performance in patients with COPD. Chest 1988; 93:977-83
13. Otte K.E., Ahlburg P., D'Amore F., Stellfels M.: Nutritional repletion in patients with emphysema. JPEN 1989; 13:152-56
14. Angelillo V.A., Bedi S., Durbes D., Dahl J., Patterson A.J., O'Donohue N.J.: Effects of low and high carbohydrate feedings in ambulatory patients with chronic obstructive pulmonary disease and chronic hypercapnia. Ann. Intern. Med. 1985; 103:883-85
15. Wilson D.O., Rogers R.M., Sanders M.H., Pennock B.E., Reilly J.J.: Nutritional intervention in malnourished patients with emphysema. Am. Rev. Respir. Dis. 1986; 134:672-77

13. Lung Alveolar Modifications and Nutrition in Malnourished Patients on Mechanical Ventilation: a Preliminary Report

S. ZANABONI, M. LUSUARDI, A. CAPELLI, S. ZACCARIA, E. L. SPADA, C.F. DONNER

Clinica del Lavoro Foundation, Institute of Care and Research Medical Center of Rehabilitation, Veruno, Novara, Italy

Introduction

That metabolic status exerts a significant influence on respiratory tract structure and function is well known. An extreme condition such as starvation in animal models can be associated with parenchymal modifications leading to atelectasis, emphysematous changes and susceptibility to infections.[1-3] In these situations a surfactant deficit has been demonstrated, related to a low uptake of precursors and/or to a decreased function of synthetic enzyme processes by type II pneumocytes.[1,2]

The role of surfactant is not limited to the preservation of an adequate airway patency, but an important involvement seems present in antibacterial defence system.[4,5] A decrease in immunoglobulin synthesis is also relevant in predisposing to lung infections in protein-caloric malnourished subjects.[6] COPD patients frequently undergo infections as a consequence of nutritional impairment and furthermore a state of immunoincompetence, evaluated with skin tests, may be a relevant feature. Among proteins, also molecules with antioxidant and antiprotease activities may result diminished from a lack of aminoacid intake.[1,2]

Patients affected by chronic respiratory failure are frequently malnourished due to both increased energy requirement and wrong dietary regimen. Malnutrition can further worsen respiratory failure by impairing functional indices and antimicrobial defences.

The pulmonary local evaluation of the cellular and biochemical correlates of malnutrition could be performed with bronchoalveolar lavage (BAL), a technique not to be employed in very compromised patients[7] unless a ventilatory support and vital function monitoring are set up. In this condition BAL can be useful for the diagnosis of infection or for the therapeutic removal of secretions. The analysis of BAL fluids may give useful information also for research purposes. In critical patients with ARDS (adult respiratory distress syndrome), for example, the analysis of alveolar cells and soluble components has revealed important information regarding pathogenesis, with potential therapeutic implications.[9,10]

Materials and Methods

We studied 7 patients with chronic respiratory failure, 6 of whom were malnourished, and 1 diabetic female affected by diffuse arteriopathy, all undergoing assisted mechanical ventilation for an episode of acute respiratory failure.

The identification data of each subject completing the study are reported in tables. Only 5 were finally considered, because in 2 COPD patients the first BAL gave very poor recoveries (< 15%), sufficient to allow microbiological analysis but not for a reliable interpretation of cytological and biochemical data.

The basic disease was COPD in 5, kyphoscoliosis in 1 and myasthenia in 1. The patient affected by diabetes and diffuse peripheral arteriopathy was studied to assess eventual major differences occurring during mechanical ventilation in absence of malnutrition and overt lung disease.

To evaluate the local pulmonary consequences of malnutrition patients were submitted to a BAL, and the alveolar cells and biochemical pattern were investigated. BAL was performed 7-10 days after intubation. All the patients were treated with total parenteral (TPN) or naso-gastric enteral nutrition (EN) or both, and the above mentioned determinations were repeated after a time interval ranging from four to six weeks in only three. Pharmacologic therapy was limited to antibiotics in case of clinically overt airway infection.

Energy requirement was calculated from the Harris-Benedict equation to evaluate the patients basal energy expenditure (BEE). We took into account a 1.5 to 2.0 value as a stress factor to correct the BEE for each patient. The goal of protein administration was to achieve a nitrogen equilibration as determined by the nitrogen balance (NB), calculated as the difference between nitrogen intake and nitrogen output (24 hour urine urea + 3) on day 1, 4, 8, 15, 22 and 30. Patients not reaching a positive NB were given an additional nitrogen load.

At the onset of nutrition we infused only the BEE requirement over the first and second days, adopting the correction factor on the 3rd day onward. The 50-55% of the infused caloric load was accounted for by fats. TPN or EN were gradually decreased when patients started to eat autonomously and stopped only when patients were able to take at least 70% of energy requirement.

To evaluate the nutritional status we determined the percentage of the ideal body weight.

Before starting TPN or EN and after four weeks we registered also the usual haematologic data.

BAL was performed during a fiber-optic bronchoscopy, passing through a naso-tracheal tube or a tracheostomic cannula and injecting a subsegment of the middle lobe or of the lingula with three 50 ml boluses of sterile saline solution at 37°C.[11]

Each aliquot was recovered and analysed separately for cellularity and cell differentials to better evaluate the bronchial component, mainly represented by the first recovery. BAL data are reported as weighed mean of the three recoveries. After a filtration of fluids on a single layer of sterile gauze, cellularity was determined with a Bürker chamber and cell differentials were evaluated on cytocentrifuge-made (Cytospin 2, Shandon, London, UK) slides stained with May-Grünwald-Giemsa.[12] Perls staining for intracytoplasmic emosiderin demonstration was performed in two cases, as specified below.

Cells were separated from supernatants with centrifugation (mod. TJ-6, Beckman, Fullerton, USA) at 4°C and 500 g for 15 minutes.

BAL supernatants were assayed for total proteins with the method of Lowry,[13] and for the main protein fractions (albumin, IgG, IgA, IgM) with an immunoturbidimetric method (ITM), using a personal modification of commercially available procedures for microalbuminuria[14] (Microalb, Urin-Pak immuno, Miles Italiana S.p.A., Ames Division, Cavenago Brianza, Italy) and blood Ig[15] (IgG Serapak, Miles Italiana S.p.A., Ames Division, Cavenago Brianza, Italy).

All BALF samples were unconcentrated. Working volumes for sample and reagents suggested by the manufacturer were modified according to the characteristics of our material: 0.3 instead of 0.1 ml and 0.7 ml instead of 1.0 ml of sample and reagent respectively were used. ITM readings were done with a spectrophotometer COMPUR M 2000 CS2 (Bayer Diagnostic-Electronic, München, FRG).

Total phospholipids were determined as phospholipidic phosphorus (P) with the Bartlett's method[16] after lipid extraction from supernatants according to Folch's procedure.[17]

Biochemical data are reported as absolute values/ml of BAL fluid and as a ratio to albumin (the protein fractions) and to total proteins (the lipid phosphorus).

Reference values are reported as mean ± standard deviation. They regard a case series of 11 healthy non smoker subjects (only 5 as for phospholipid evaluation).

Results

BAL did not cause major effects because all the subjects were submitted to assisted mechanical ventilation during the whole procedure.

The recoveries were good in restrictive patients (Tables IV and V), comparable to those obtainable in normal case series, while obstructive disease was always associated with a great loss (70-90%) of the infused saline. In this case, BAL fluids

Table I

| Patient: M.A. | sex: F | age: 64 | smoking habit: non-smoker |

diagnosis: acute respiratory failure in diabetic arteriopathy and congestive heart failure.

Ideal Body Weight (%)	108.000
Ventilation	IPPV
BUN (mg/dl)	29.000
Creatinine (mg/dl)	0.960
WBC/mm^3	9.500
Hb (g/dl)	12.600
Total proteins (g/dl)	6.500
Intake (kcal/day)	BEE x 0.71

BAL data:	BAL	Reference values
vol. recovery %	60.000	63.30±12.70
cells/ml x 10^3	190.200	185.00±53.40
macrophages %	94.700	89.00± 4.60
lymphocytes %	1.050	9.70± 4.10
neutrophils %	4.090	0.90± 0.80
eosinophils %	0.080	0.32± 0.30
basophils %	0.110	< 0.5
viability %	88.000	> 95
total proteins (mg % ml)	6.300	10.80±2.80
albumin (mg %ml)	2.860	2.97±0.97
IgG (mg % ml)	0.090	0.33±0.16
IgG/alb	0.030	0.12±0.09
IgA (mg % ml)	0.150	0.08±0.04
IgA/alb	0.050	0.03±0.01
IgM (mg % ml)	0.098	< 0.05
IgM/alb	0.030	< 0.02
total phospholipidic P/ total proteins (µg/ mg)	8.900	13.97±3.94

Table II

| *Patient:* A.M. | *sex:* M | *age:* 73 | *smoking habit:* ex-smoker |

diagnosis: Exacerbation of chronic respiratory failure in COPD

Ideal Body Weight (%)	89.000	
Ventilation	IPPV	
BUN (mg/dl)	55.000	
Creatinine (mg/dl)	0.810	
WBC/mm^3	15.200	
Hb (g/dl)	14.400	
Total proteins (g/dl)	5.900	
Intake (kcal/day)	poor ?	
BAL data:	BAL 1	Reference values
vol. recovery %	16.000	63.30±12.70
cells/ml x 10^3	353.600	185.00±53.40
macrophages %	38.400	89.00± 4.60
lymphocytes %	0.300	9.70± 4.10
neutrophils %	71.000	0.90± 0.80
eosinophils %	0.200	0.32± 0.30
basophils %	0.200	< 0.5
viability %	88.000	> 95
total proteins (mg % ml)	12.800	10.80±2.80
albumin (mg % ml)	0.280	2.97±0.97
IgG (mg % ml)	1.800	0.33±0.16
IgG/alb	1.410	0.12±0.09
IgA (mg % ml)	1.200	0.08±0.04
IgA/alb	0.940	0.03±0.01
IgM (mg % ml)	0.203	< 0.05
IgM/alb	0.160	< 0.02
total phospholipidic P/ total proteins (μg/mg)	3.300	13.97±3.94

were shown to be prevalently bronchial in origin, with a very high percentage of neutrophils and bronchial epithelial cells. Microbiological quantitative cultural analysis led to the isolation of pseudomonas aeruginosa in two COPD patients at a Colony Forming Unit (CFU) > 10^4.

The data from the two restrictive (Tables IV and V) and two among the obstructive patients (Table II and III) will be considered in detail and shown as case reports along with the case represented by the diabetic subject (Table I).

In this patient, who had no history of lung disease, BAL recovery was comparable to normal along with total cellularity and cell differentials, the only exceptions being a mild neutrophilia (4%), prevalent in the first recovery and a low lymphocyte percentage (1%). As for total proteins they were moderately lower than normal, while, among fractions, albumin, IgA and IgM were comparable to the normal levels and IgG absolute quantity showed a decrease in comparison to control value but with a normal IgG/Alb. Total phospholipids were slightly lower than normal. A control of the alveolar characteristics was not possible after a 4 week follow up period with TPN, because the patient died from massive mesenteric thrombosis.

In the two COPD patients we were able to study in more depth (Tables II and III), an airway infection was relevant at baseline, with neutrophils being very high as both percentage and absolute values, while lymphocytes tended to be lower than normal. Total proteins and albumin were in the normal range and this stands for an absence of bronchial exudation and/or blood/alveolar barrier permeability increase at the moment of BAL. On the contrary a level of immunoglobulins higher than normal was present, probably related mainly to a local synthesis, since not only absolute but also albumin standardized data were parallelly increased. Total phospholipids were markedly reduced.

In only one of the two patients a second lavage was possible after four weeks and showed a relevant decrease of cellularity and particularly of neutrophil number (but with a percentage increase) along with a marked decrease of Ig/albumin level but without any improvement of phospholipid concentration (Table III).

In restrictive patients recoveries were relatively good, with amelioration of the second versus the first BAL. Basal BAL was associated with a high yield of red blood cells, without any visible sign of haemorrhage at the bronchoscopic examination. Red blood cells were not observable within the macrophage cytoplasm and the Perls' reaction was negative, thus demonstrating an acute extravasation of blood across the blood-alveolar barrier and/or mucosal capillaries. Total cellularity was normal in one case and markedly increased in the second, with cell differentials characterized by slight lymphocytosis and neutrophilia in the former case and a marked increase of granulocytes in the latter one. The basal alteration of biochemical parameters was similar in both cases, with a relevant augmentation of total proteins, albumin and Ig fractions. Concerning immunoglobulins, it is necessary to underline that Ig to albumin ratio was only moderately increased, thus revealing an

Table III

Patient: A.L.	*sex:* M	*age:* 70	*smoking habit:* ex-smoker

diagnosis: Chronic respiratory failure post chest trauma pneumonectomy, diffuse lung tuberculosis outcomes and chronic bronchitis.

	5 week interval		
Ideal Body Weight (%)	98.000	98.000	
Ventilation	IPPV/spont.	IPPV/spont.	
BUN (mg/dl)	47.000	41.000	
Creatinine (mg/dl)	0.800	0.800	
WBC/mm^3	8.700	9.700	
Hb (g/dl)	11.700	10.900	
Total proteins (g/dl)	6.000	5.600	
Intake (kcal/day)	BEE x 1.65		
BAL data:	BAL 1	BAL 2	Reference values
vol. recovery %	30.000	25.000	63.30±12.7
cells/ml x 10^3	690.000	202.000	185.00±53.4
macrophages %	36.800	14.400	89.00±4.60
lymphocytes %	4.900	1.100	9.70±4.10
neutrophils %	57.100	83.500	0.90±0.80
eosinophils %	1.300	1.040	0.32±0.30
basophils %	-	0.050	< 0.5
viability %	90.000	92.000	> 95
total proteins (mg % ml)	15.600	13.000	10.80±2.80
albumin (mg % ml)	4.200	4.400	2.97±0.97
IgG (mg %ml)	4.940	1.450	0.33±0.16
IgG/alb	1.180	0.330	0.12±0.09
IgA (mg % ml)	1.050	0.580	0.08± 0.04
IgA/alb	0.250	0.130	0.03±0.01
IgM (mg % ml)	0.158	0.112	< 0.05
IgM/alb	0.038	0.025	< 0.02
total phospholipidic P/ total proteins (µg/mg)	3.200	2.4	13.97±3.94

Ig increase of prevalently haematogenous origin. The phospholipidic phosphorus was lower than normal but with a difference of slight relevance.

A control of BAL data after six weeks showed a cellularity increment in the patient with basal normal value (Table IV), concomitant with an improvement of clinical and nutritional status. An absolute and percent increase of lymphocytes was more evident than in the baseline situation, but neutrophils declined within normal values. Cellularity and granulocytes decreased in the second case (Table V) after four weeks but remaining markedly higher than normal, with a relevant increase of lymphocytes. This patient showed a slight amelioration of nutritional status, but a persistence of blood neutrophilia was seen in concomitance of the sign of alveolar inflammation.

Red blood cells were almost absent in the first case, while present but significantly decreased in the second one.

Concerning biochemical data, in both cases total proteins, albumin and Ig fractions decreased in comparison with baseline absolute values, with Ig/albumin not relevantly different, but with a persistence of alterations versus control data. No particular modifications of total phospholipids were shown in the tested case.

Discussion

Local alveolar consequences of malnutrition are difficult to assess in man for both ethical and methodological problems. Malnourished patients admitted to our *Respiratory Intensive Care Unit* are usually affected by chronic respiratory failure with inflammatory exacerbation. They may represent a possible subject of investigation, provided that our means of study can be adequate for both reliability and applicability.

BAL has proved to be a very good technique in the study of terminal airways, concerning cytological and biochemical aspects,[9,10] but critical conditions have always been considered as an absolute contra-indication.[7,9,10] When patients are intubated and submitted to mechanical ventilation these problems can be overcome, since respiratory function can be continuously monitored and supported. Using a "T"-piece connection (Mount catheter), a fiber-optic bronchoscope can be easily inserted through the naso (oro) tracheal tube without stopping ventilation, and the tip incuneated in the bronchial branch of the lung area to be studied. A clear benefit of the examination should obviously be considered before planning a broncho-alveolar lavage in a critical patient, such as the necessity of a microbiological study.[8]

For these reasons few malnourished patients have been studied for alveolar characteristics until now at our Division. What we discuss in this paper can be considered only as a series of case reports, whose interest is mainly limited to the demonstration of new possibilities of cytological and biochemical investigation in

Table IV

| Patient: F.R. | sex: F | age: 40 | smoking habit: non-smoker |

diagnosis: Acute respiratory failure in kyphoscoliosis

	6 week interval	
Ideal Body Weight (%)	80.600	83.3
Ventilation	IPPV	spontaneous
BUN (mg/dl)	48.000	28
Creatinine (mg/dl)	0.700	0.61
WBC/mm^3	9.700	8.700
Hb (g/dl)	11.500	11.2
Total proteins (g/dl)	5.700	6.2
Intake (kcal/day)	BEE x 1.740	

BAL data:	BAL 1	BAL 2	Reference values
vol. recovery %	42.00	58.000	63.30±12.70
cells/ml x 10^3	240.30	593.600	185.00±53.40
macrophages %	76.40	79.200	89.00±4.60
lymphocytes %	17.60	19.500	9.70±4.10
neutrophils %	5.30	0.500	0.90±0.80
eosinophils %	0.71	0.760	0.32±0.30
basophils %	-	-	< 0.5
red blood cells	numerous	rare	absent
viability %	88.00	90.000	> 95
total proteins (mg % ml)	46.70	29.20	10.80±2.80
albumin (mg % ml)	18.00	11.400	2.97±0.97
IgG (mg %ml)	5.17	2.590	0.33±0.16
IgG/alb	0.29	0.230	0.12±0.09
IgA (mg % ml)	2.10	1.560	0.08± 0.04
IgA/alb	0.12	0.140	0.03±0.01
IgM (mg % ml)	2.30	0.083	< 0.05
IgM/alb	0.13	0.007	< 0.02
total phospholipidic P/ total proteins (μg/mg)	6.60	5.800	13.97±3.94

the nutritional field in man.

The first case (a well nourished patient without overt lung disease, Table I) revealed an almost normal airway situation. A mild neutrophilia, in the first recovery only, stands for bronchial inflammation, probably due to the invasive resuscitation procedures. Biochemical parameters demonstrate a completely normal blood-alveolar barrier permeability. This particular is interesting if compared with the situation found in both the two restrictive malnourished patients (Tables IV and V), characterized by a marked fragility/hyperpermeability of alveolar and/or mucosal capillary barrier, evidenced by a high presence of red blood cells in BAL fluids, along with high levels of proteins without a parallel increase of protein fractions to albumin ratios. Probably suction manoeuvres during BAL, though gently performed with a syringe and manual aspiration, were responsible for blood extravasation from alveoli/mucosa. No blood was visible at the bronchial level in correspondence to the bronchoscope tip, thus bronchoscopy should not be responsible per se. An accurate microscopic examination of macrophages stained with both May-Grünwald-Giemsa and Perls (always negative) permitted us to exclude a chronic alveolar leakage of blood. In one case (Table V) there were marked signs of inflammation (very high neutrophilia), not manifest in the second one (Table IV), but both cases presented with superimposable alterations of biochemical data.

In parallel with the improvement of clinical and nutritional status, more evident in the second case (Table IV), we observed after 4 and 6 weeks respectively, a reduction and a disappearance of red blood cells along with a comparable diminution of protein absolute values, without modification of albumin standardized data. Protein values being still higher than normal, we can suppose an only partial reduction in capillary hyperpermeability. It would be interesting to investigate how much time is needed, in appropriate nutritional conditions, to restore blood-alveolar barrier function to normal. Such fragility is probably evidenced, at least in part, when BAL is performed and it is thus difficult to argue if it may cause particular problems per se. It would be possible, for example, that a leakage of blood proteins might impair the function of surfactant with a reduction of lung compliance. It has been in fact demonstrated that surfactant can be inactivated by fibrinogen.[6] In our cases phospholipids did not seem greatly reduced but we are not able to discriminate between lung and serum phospholipids nor could we measure tension-active properties of BAL fluids.

Another interesting feature found in the restrictive patients was an absolute and percentage increase of lymphocytes in the second versus the first BAL. This probably means an improvement in local defences.

COPD patients usually do not allow a correct BAL performance, because very poor recoveries are usually obtained, as happened in three among our cases, who are reported here. These fluids are mainly bronchial in origin and are in any case suitable for microbiological analysis.

Table V

Patient: G.P.	sex: F	age: 45	smoking habit: non-smoker

diagnosis: Acute respiratory failure after abdominal surgery in myasthenia

	4 week interval		
Ideal Body Weight (%)	73.000		75.000
Ventilation	IPPV		IPPV
BUN (mg/dl)	92.000		54.000
Creatinine (mg/dl)	0.680		0.370
WBC/mm^3	17.100		15.400
Hb (g/dl)	9.200		11.500
Total proteins (g/dl)	6.700		8.200
Intake (kcal/day)	BEE x 1.920		

BAL data:	BAL 1	BAL 2	Reference values
vol. recovery %	34.000	43.000	63.30±12.70
cells/ml x 10^3	1251.200	844.600	185.00±53.40
macrophages %	32.700	50.600	89.00± 4.60
lymphocytes %	5.600	14.600	9.70± 4.10
neutrophils %	59.800	33.600	0.90± 0.80
eosinophils %	1.720	0.700	0.32± 0.30
basophils %	0.100	0.050	< 0.5
red blood cells/ml x 10^6	4.200	1.200	absent
viability %	90.000	92.000	> 95
total proteins (mg % ml)	58.600	25.900	10.80± 2.80
albumin (mg % ml)	22.630	7.970	2.97± 0.97
IgG (mg %ml)	9.180	5.650	0.33± 0.16
IgG/alb	0.410	0.710	0.12± 0.09
IgA (mg % ml)	2.540	1.350	0.08± 0.04
IgA/alb	0.110	0.170	0.03± 0.01
IgM (mg % ml)	0.370	0.110	< 0.05
IgM/alb	0.016	0.014	< 0.02

In two cases a certain aliquot was recovered, enough to permit a cytological and biochemical analysis. The main feature was a high neutrophilia and a marked increase of absolute and albumin standardized Ig values, which stand for a relevant local immunoglobulin synthesis. In the case reported in table II it was not possible to perform a second BAL. The patient of table III was submitted again to BAL after 5 weeks. A reduction of cellularity and Ig was found, but with a persisting high neutrophilia, despite a good nutritional status at baseline, well maintained with nutritional supplementation. The persistence of active signs of inflammation in airways even in steady state clinical conditions is probably one of the factors that renders the prognosis of mechanically ventilated COPD patients worse than that of patients without relevant lung parenchymal involvement.

From our preliminary data we can thus conclude that in malnourished restrictive patients without marked parenchymal involvement BAL can be performed with a good yield of fluids, whose analysis can reveal useful information on the local effects of malnutrition itself and nutritional therapy.

In particular some points can be underlined as possible items worth being investigated. What is the functional meaning of the lymphocyte increase in mechanically ventilated malnourished restrictive patients (without gross lung involvement) recovering from an episode of acute respiratory failure?

What is the basic substrate of the fragility/hyperpermeability of the capillary barrier? Is there a functional surfactant deficiency? And finally, what is the actual role of nutritional intervention in modifying these characteristics?

On the contrary, BAL does not seem a suitable method to assess the alveolar effects of malnutrition and nutritional therapy in COPD patients. It can however be a reliable tool to assess lung infections in critical subjects.

References

1. Sahebjami H.: Nutrition and lung function. In: D. Massaro (Ed.) *Lung cell biology*. New York, Marcel Dekker Inc. 1989; pp. 1247-1278
2. Wilson D.O., Rogers R.M., Openbrier D.: Nutritional aspects of chronic obstructive pulmonary disease. Clinics in Chest Medicine 1986; 7 (4):643-656
3. Rochester D.F., Esau S.A.: Malnutrition and the respiratory system. Chest 1984; 85:411-415
4. Jarstrand C.: Role of surfactant in the pulmonary defence system. In: Robertson B., Van Golde L., Batenburg J. (Eds.) *Pulmonary surfactant*. Amsterdam, Elsevier 1984; 187-201
5. O'Neill S.J., Lespearance E., Klass D.J.: Human lung lavage surfactant enhances staphylococcal phagocytosis by alveolar macrophages. Am. Rev. Respir. Dis. 1984; 130:1177-1179
6. Barrett T.A., Robin A.P., Armstrong M.K., Bone R.C.: Nutrition and respiratory failure. In: Bone R.C., George R.B., Hudson L.D. (Eds.) *Acute Respiratory Failure*. New York, Churchill Livingstone 1987; pp. 265-303
7. Report of the European Society of Pneumology Task Group on BAL, Klech H., Pohl W. (Eds.) Technical recommendations and guidelines for bronchoalveolar lavage (BAL). Eur. Respir. J. 1989; 2:561-585

8. Torres A., De La Bellacasa J.P., Xaubet A., Gonzalez J., Rodriguez-Roisin R., Jimenez De Anta M.T., Agusti Vidal A.: Diagnostic value of quantitative cultures of bronchoalveolar lavage and telescoping plugged catheters in mechanically ventilated patients with bacterial pneumonia. Am. Rev. Respir. Dis. 1989; 140:306-310

9. Rossi G.A.: Bronchoalveolar lavage in the investigation of disorders of the lower respiratory tract. Eur. J. Respir. Dis. 1986; 69:293-315

10. Reynolds H.Y.: Bronchoalveolar lavage. Am. Rev. Respir. Dis. 1987; 135:250-263

11. Velluti G., Capelli O., Lusuardi M., Braghiroli A., Pellegrino M., Milanti G., Benedetti L.: Bronchoalveolar lavage in the normal lung. First of three parts: protein, enzymatic and ionic features. Respiration 1983; 44:403-410

12. Velluti G., Capelli O., Lusuardi M., Braghiroli A., Azzolini L.: Bronchoalveolar lavage in the normal lung. Second of three parts: cell distribution and cytomorphology. Respiration 1984; 46:1-7

13. Lowry O.H., Rosebrough N.J., Farr A.L., Randall R.J.: Protein measurement with the Folin phenol reagent. J. Biol. Chem. 1951; 193:265-275

14. Donà V., Maierna M., Tarenghi G., Berti G.: An immunoturbidimetric assay for microalbuminuria. Ann. Clin. Biochem. 1987; suppl. 2: 272-273

15. Donà V., Papagni M., Tarenghi G., Aguzzi F., Berti G.: Specific serum protein quantitation by a simple immunoturbidimetric method. G. Ital. Chim. Clin. 1987; 12 (3):205-214

16. Bartlett G.R.: Phosphorus assay in column chromatography. J. Biol. Chem. 1959; 234:466-468

17. Folch J., Lees M., Sloane-Stanley G.H.: A simple method for the isolation and purification of total lipides from animal tissues. J. Biol. Chem. 1957; 226:497-509

14. Nutrition in A.R.D.S. and Multiorgan Failure

J. M. KINNEY

Visiting Professor and Physician, Rockefeller University; Attending in Medicine and Surgery, St Luke's-Roosevelt Hospital Center, New York, USA

The optimum nutrition for any individual depends not only upon their body size, sex and age but also upon their underlying metabolic state. This is particularly true for patients needing specialized nutritional support in the hospital. Such patients commonly fall into three categories: the acutely ill or injured patient with accelerated catabolism, the seriously depleted patient, and the patient with compromise of a vital organ system. Patients in each category may suffer from ventilatory loss, if not outright ventilatory failure. The role of nutrition in the management of the patient with loss of ventilatory function has received growing attention in the past decade. No other clinical situation demonstrates more clearly the importance of adjusting the nutritional intake to achieve specific objectives while avoiding the side effects of improper or overenthusiastic nutrition.

Parallel Growth of Ventilatory Care and TPN - 1965 to 1980

Striking similarities can be noticed in the timing, circumstances and parallel development of ventilatory care and nutritional support over the last two decades.[1] A major stimulus for developing ventilatory care and intensive care in general was the evolution of mechanical ventilation. The introduction of total parenteral nutrition played a similar role in stimulating the concept of nutritional support.

The incidence of respiratory paralysis from poliomyelitis stimulated the development of tank respirators which were often grouped in the hospital in order to provide for efficient care. The introduction of blood gas electrodes in the 1960's provided the necessary diagnostic and monitoring capability for intelligent ventilatory support. Physicians began to recognize a new syndrome of Adult Respiratory Distress Syndrome (ARDS) which was distinct from the ventilatory failure previously recognized with advanced chronic lung disease.[2] Patients with this new syndrome required medicalcare which was difficult to deliver in the tank respirators from the polio era, thus stimulating the introduction of successive generations of mechanical devices for providing intermittent positive pressure to the airway. Such ventilatory support could only be provided in an intensive care setting with properly trained staff. The 1970's saw increasing sophistication in ventilatory intensive care, partly as the result of increasing clinical awareness of early ventilatory compromise, the availability of a skilled team in the intensive care unit, and improved procedures and devices which minimized the risks of airway infection, inadequate humidification and improper use of airway pressure.

During the 1950's the only common intravenous nutrition provided 5% dextrose and water, more commonly given as a source of sodium-free water than as a form of nutrition. Investigators in both the U.S. and Europe were exploring the use of protein hydrolyzates by vein but there was a widespread feeling that the provision of nitrogen would never improve nitrogen balance until a safe and effective lipid source could allow an adequate calorie intake. Various i.v. lipids were tested before an emulsion developed by Wretlind in the early 1960's was shown to be safe and effective, however it was not available in the U.S. until nearly 15 years later. With no lipid preparation available in the U.S., Dudrick introduced the use of "hyperalimentation" which involved the continuous administration of hypertonic glucose and amino acid with vitamins and minerals into a central vein. The concept of total parenteral nutrition stimulated the development of techniques for nutritional assessment, and attention to the incidence of hospital malnutrition during the 1970's. New products improved the safety and availability of TPN, and at the same time encouraged better use of enteral nutrition whenever possible. Just as intensive care has undergone specialization to provide particular care for selected types of patients, specialized solutions are being developed for optimum support of patients, specialized nutritional problems associated with organ failure, whether hepatic, renal or pulmonary in origin.

Ventilatory Capacity and Metabolic Need - A Critical Balance

Both metabolic "demand" and ventilatory "supply" are modified by acute illness and injury, and both are also modified by weight loss and tissue depletion. Body composition becomes altered in both acute catabolism and chronic depletion, in ways which may have undesirable influences on ventilation. Both metabolic rate and minute ventilation have often been expressed as a function of body surface area or body weight alone. The fat-free mass,[3] or the body cell mass,[4] would appear to be more useful indicators of the metabolic body size than either body weight or surface area, however such measurements are not widely available in the clinical setting.

The measurement of resting energy expenditure (REE) in hospital patients had been restricted to clinical research until the availability of bedside devices for measuring gas exchange allowed this measurement to become part of clinical management. Some of these devices which provide information on gas exchange also provide information on the pattern of ventilation, although the correlation between minute ventilation and either VO_2 or VCO_2 has not been adequately studied in the clinical setting.

Measurement of the REE of patients during the 1960's revealed relatively large increases in acutely catabolic patients and significant decreases in patients with weight loss and advanced tissue depletion. The uncomplicated postoperative patient has an REE which varies from zero to 10% above pre-operative levels, the multiple injury patient shows increases of 10 to 25% for two to three weeks, the patient with major sepsis is usually 20 to 50% above predicted normal at the height of the febrile response.

The REE of the major burn in the period between 1965 and 1975 was 50 to 100% above predicted levels.[5] This severe hypermetabolism was the result of prolonged open treatment with antibacterial ointments in an inadequately warmed environment, plus continued catabolic stimulus from dead tissue. Now the range of REE in the major burn has been reduced to the range of 25 to 60% above normal by treatment in a warmed environment with early excision and covering with either grafted skin or other effective substitute which controls the influence of dead tissue, reduces the risk of burn wound sepsis and prevents the obligatory evaporative cooling through the full thickness burn wound. At the same time, it has become rare to see the patient with advanced tissue depletion and REE which is depressed from 20 to 35% below predicted normal. Therefore, a cross section of patients in a modern hospital will tend to have measurements of REE which vary from plus 40% to minus 20%, rather than the plus 100% to minus 35% which was observed in the 1960's.

The energy expenditure during fever in a variety of medical conditions was reported by DuBois[6] and later compared with surgical conditions by Kinney et al.[7,8] Mean values for the caloric equivalent of fever were similar in the medical and surgical conditions. However, surgical patients who required reoperation after a period of undernutrition and weight loss showed a decrease in the caloric equivalent of fever which was similar to that previously shown for the depleted patients with chronic pulmonary tuberculosis. The surgical counterpart of acute typhoid fever was found to be severe peritonitis with an increased caloric equivalent of fever and nitrogen loss.

Macklem et al.[9] have examined the pathophysiology of inspiratory muscle fatigue in relation to oxygen consumption. They emphasized that the critical value of the rate of energy consumption of the inspiratory muscles beyond which fatigue occurs, appears to be predictable in some instances from the relationship between energy demands and energy supplies, rather than from the percentage of fatigue-resistant fibers in the inspiratory muscles. The normal O_2 cost of breathing is generally considered to be in the range of 2 to 5% in the normal resting adult. Macklem and co-workers showed that the O_2 demands of the inspiratory muscles may equal the entire O_2 supply to the whole body doing fatiguing inspiratory work.

Harpin et al[10] studied 20 patients who were mechanically ventilated for critical illness in the intensive care unit. A positive correlation was shown between the oxygen cost of breathing and the total time in days before the patient could be weaned from the ventilator. This is consistent with the plea of Gilston[11] to place new attention on ARDS to improved means of lessening the O_2 consumption.

Acute Ventilatory Failure (ARDS)

An interval of nearly 20 years has passed since the term "Adult Respiratory Distress Syndrome" was introduced.[2] It was used to describe the combination of tachypnea, decreased compliance, hypoxaemia and evidence of interstitial pulmonary edema on x-ray, in a patient without prior lung disease. Experience in the Vietnam war caused it to be known as "shock lung" or "post-traumatic pulmonary insufficiency".[12] With more clinical experience and experimental studies, the low flow state from haemorrhage has seemed less likely to result in ARDS unless resuscitation required massive blood transfusion. The most common aetiology was thought to be related to infection or a hyperdynamic condition which suggested infection.

Originally ARDS was defined as an acute respiratory failure secondary to a non-pulmonary cause and associated with non-cardiogenic pulmonary edema.[12] However, the list of possible aetiological factors has grown dramatically. Postoperative patients with ventilatory failure were often lumped together regardless of whether pneumonia, massive atelectasis or pulmonary emboli were involved at the onset. More recently the risk factors have been considered as direct or indirect:

Direct: Aspiration
            Pulmonary contusion
            Smoke inhalation
            Near drowning
Indirect: Endothelial Injury from:
            Systemic sepsis
            Distant injury
            Pancreatitis
            Fat embolism (major fx)

The mechanisms and mediators of the acute catabolic response and possible ventilatory compromise were initially considered to involve the central nervous system or conventional hormones. The past decade had seen an explosion of additional factors which may be involved singly, or together, to produce local, regional or systemic responses.[13-16]

CNS	Systemic
Hormones	Responses
Lymphocytes	
Monocytes (circ. macrophages)	_____
Tissue Macrophages - cytokines	
Endotoxin	Regional or
Prostaglandins	Local
Histamine	Responses
Kinins & Enzymes	

Acute pancreatitis is often associated with pulmonary dysfunction and approximately 10% of such patients will develop full blown ARDS. Pancreatitis is thought to activate the complement system causing pulmonary leukostasis and triggering the coagulation system to release kinins. Oxygen free radicals released from activated neutrophils are thought to be an important mechanism for producing endothelial injury.

Immune complexes containing immunoglobulins appear to cause a complex series of humoral and cellular reactions leading to acute lung injury. Oxygen free radicals may also damage the pulmonary endothelium from the activation of tissue macrophages.

Arachidonic acid normally exists bound to phospholipids of cellular membranes including neutrophils. These phospholipids may be released when cellular injury occurs and when, by interacting with phospholipase, arachidonic acid is released which serves as substrate for many powerful vasoactive substances. Prostanoids utilizing the cyclo-oxygenase pathway include both vasoconstrictors and vasodilators, while leukotrienes exert strong chemotactic properties which may result in increased permeability.

One unified hypothesis of cellular and humoral events leading to ARDS suggests the following:

1. One or more risk factors activates complement.
2. Complement stimulates chemotaxis and causes pulmonary leukostasis.
3. Neutrophil release of oxygen radicals and enzymes damages pulmonary endothelium with increased permeability.
4. Leukotrienes also increase permeability and attract neutrophils.
5. Histamine from trapped platelets yields microthrombi.

6. Increased interstitial pressure and edema cause terminal airway narrowing and alveolar collapse.

The current management of ARDS, despite many valuable approaches, remains supportive. There is no proven therapy which will control or reverse the etiologic processes, although various experimental therapies are under study. Positive end-expiratory pressure has been widely used but it is not clear whether its early use can abort the development of ARDS. Corticosteroid use remains controversial but the majority of recent reports have indicated no improvement in ARDS.

The survival rate reported in recent studies has remained near 50%, which is relatively unchanged since ARDS was first described. The most widely accepted definition of ARDS is hypoxaemia with a PaO_2 less than 60 mmHg while the FiO_2 is over 0.6 at a time when the pulmonary artery wedge pressure is less than 12 cm of water. The strict application of these criteria will define a severe subset of patients with a high mortality and may fail to recognize patients in the early stages of ARDS where future therapy may be the most effective.[17]

Multiple Organ Failure (MOF)

This is a syndrome which is usually initiated by a severe physiologic insult, such as multiple trauma or severe infection with septic shock. Resuscitation is followed by a brief period of clinical stability and then fever, hypo-albuminaemia and some aspect of cardiopulmonary failure appear. Initial cardiopulmonary support appears to be successful despite abnormal chemical evidence of hepatic dysfunction. The patient is often considered to have infection although localized sepsis can be found and there may not even be a positive blood culture.[18]

The precise causes and mechanisms of ARDS are poorly understood in man and despite supportive treatment with mechanical ventilation and diuretic, more than half of patients with ARDS die. The mortality of ARDS was found to be far higher than previously realized in a collaborative study by NIH in 1979 involving 686 adult patients in nine medical centres.[19] Mortality up to age 65 was 41% with ARDS alone while over the age of 65 the death rate rose to 68%. The findings were similar to those of Knaus et al[20] in that the mortality rose with duration and with each successive organ failure. The mortality of patients with severe injury or infection has decreased significantly in recent years in association with the support of vital organ function which is possible in modern intensive care units.

In spite of the best current therapy, some patients undergo progressive deterioration of organ function. Thus MOF is presently the main cause of death in most intensive care units. Sepsis has usually been incriminated as the cause of MOF. However patients may demonstrate the signs and symptoms of generalized inflammation without a septic focus or a positive blood culture. Goris et al.[21] published an extensive study of MOF and concluded that sepsis is probably not the essential cause. Instead they offered the hypothesis that massive activation of inflammatory mediators by severe tissue injury or intra-abdominal sepsis resulted in systemic damage to vascular endothelium causing increased permeability and edema with impaired oxygen availability to the mitochondria despite adequate oxygen transport. They named the sequence "generalized autodestructive inflammation".

Endotoxin has been reported to produce widespread pathophysiologic changes, but its relation to the acute catabolic response to injury and infection has not been clearly defined. However, the suggestion has been made that many of the changes due to endotoxin are similar to those associated with the inflammatory peptide, tumor necrosis factor or cachectin. A hypermetabolic state exists with ARDS which has lead Breedle et al.[25] to examine whether endotoxin may influence the critical oxygen delivery to skeletal muscle in conditions such as ARDS. They reported that endotoxin caused a small but definite O_2 extraction defect in muscle (the threshold O_2 delivery below which VO_2 falls). These workers found that this was not corrected by increasing arterial oxygen levels and suggested that this might explain the hypermetabolism of ARDS.

Nutrition in ARDS and MOF

A traditional goal of nutritional support for a patient requiring mechanical ventilation has been to restore adequate strength to the respiratory muscles. This is thought to require an aggressive nutritional regimen with a strongly positive balance of calories and nitrogen. Unfortunately such nutrition may also increase metabolic demand, thereby increasing the respiratory workload and ultimately result in clinical deterioration.[22-24] An acutely ill patient with a tendency toward hyperventilation will have this tendency increased by aggressive nutritional support. Some of this tendency increased by aggressive nutritional support. Some of this is presumably related to the mechanisms underlying diet-induced thermogenesis which may involve the brain and/or peripheral neuroendocrine effects. The possible relation of hormones, lymphokines and prostaglandins to tissue energy expenditure and hence a greater need for ventilation is the subject of active study. The balance of ventilation and metabolism in acute catabolic states has been studied by simulating acute catabolism by a 5 hour infusion of cortisol, glucagon and epinephrine in normal subjects.[26] Ventilation was uniformly increased to a greater degree than either $\dot{V}O_2$ or $\dot{V}CO_2$. Ventilation at rest and during breathing low levels of CO_2 was associated with significant decreases in arterial CO_2 tension.

High carbohydrate loads raise the respiratory quotient by increasing the CO_2 production which stimulates ventilation. This is particularly noted when the carbohydrate intake approaches the resting energy expenditure, presumably due to the high RQ (high $\dot{V}CO_2$) associated with lipogenesis. High glucose, lipid-free,

TPN causes an increased norepinephrine output.

Amino acid infusions have been shown to restore a depressed ventilatory drive after 7 days of hypocaloric nutrition in normal subjects and in chronically depleted patients. Further studies are needed to determine how much of this nutrient influence on ventilation is direct, perhaps by influencing neurotransmitter activity in the respiratory centre, and how much is secondary to some systemic effect which modifies the diet-induced thermogenesis associated with amino acids.

Intravenous lipid emulsions are of established importance as a ready source of calories and to prevent essential fatty acid deficiency. However, the relation of lipid intake to surfactant synthesis, membrane composition and prostaglandin metabolism is not well established although potentially of great importance.

It is of the greatest importance to establish proper objectives when undertaking nutrition for the patient with ARDS. These objectives must then be reviewed and perhaps modified with the onset of MOF. Nutritional support has traditionally been provided with the intent to achieve a positive balance of calories and nitrogen. The primary reason for establishing a positive balance of calories and nitrogen is to build tissue. Building tissue is seldom the first priority for the patient with ARDS, and particularly with MOF. Such patients should be given enough intake for equilibrium and hopefully be weaned from mechanical ventilation before undertaking the long, slow process of tissue synthesis.

Proper nutrition during the weaning process requires judgement and flexibility. Weaning is usually beneficial to the patient's convalescence, after which the patient will hopefully move toward increasing dietary intake by normal means. Weaning is usually a matter of hours to a few days, during which time the daily intake may need to have the total amount of calories and nitrogen reduced to one half, or less, with the carbohydrate to fat ratio lowered to yield an RQ of approximately 0.75 (less CO_2).

An exception to the above approach relates to the occasional patient who can be weaned from the ventilator with sufficient skill, but remains off the ventilator in such a marginal state that efforts to establish a rising nutritional intake produce a return of ventilatory failure. Such a patient would have been better treated had he remained on mechanical ventilation while dietary intake was carefully raised to an anabolic level and weaning delayed until ventilatory improvement was achieved.

Hypophosphataemia with glucose loads can result in disorientation and decreased respiratory muscle work. The condition can be treated by phosphate supplementation and by reducing the glucose intake.

Organ failure is often associated with sodium and water retention, which can be insidious and cannot be recognized merely by observing urine outputs. Calibrated bedscales in the hands of a trained ICU staff can alert one to the onset of undesired weight gain which is as much of a "red flag" as the onset of hyperglycemia on a given glucose intake. Both of these are indications of an inflammatory development which may appear before a significant febrile response. Improper use of nutritional support can contribute to the tendency toward sodium and water retention which may complicate ventilatory management and delay weaning from the ventilator.

There is increasing attention to the function of the intestinal mucosal barrier in preventing the translocation of bacteria and endotoxin into the portal circulation with activation of platelets and white cells either in the liver or in the blood returning from the right heart to the lung. The old nutritional maxim, *If the gut is available - use it*, is being extended to the idea that if the gut fails to receive specific luminal nutrients, a loss of mucosal barrier function can occur.[27]

Thus, the role of glutamine, short chain fatty acids and ketones as specific fuel for the intestinal mucosa may be of major importance in protecting other organs, particularly the lungs, from failure due to inflammatory damage to the vascular endothelium.

References

1. Kinney J.M.: Nutrition in the intensive care patient. Critical Care Clinics, January 1987; 3, 1: 1
2. Ashabaugh D.G., Petty T.L., Bigelow D.B.: et al.: Acute respiratory distress in adults. Lancet 1967; 2:319-325
3. Grande F.: Body weight, composition and energy balance. In: Olson RE (Ed) *Present Knowledge in Nutrition*. 5th ed. The Nutrition Foundation, Washington DC, 1984; 7
4. Moore F.D., et al.: The Body Cell Mass and Its Supporting Environment. Philadelphia, W.B. Saunders Company, 1963
5. Kinney J.M., Duke J.H.Jr., Long C.L., Gump F.E.: Tissue fuel and weight loss after injury. J. Clin. Path. 23 Suppl. (Roy. Coll. Path.) 1970; 4:65-72
6. DuBois E.F.: Fever. In: *Basal Metabolism in Health and Disease*. Philadelphia, Lea & Febiger, 1924; p. 311
7. Kinney J.M., Roe C.F.: The caloric equivalent of fever. I. Patterns of postoperative response. Ann. Surg. 1962; 156:610-622
8. Roe C.F., Kinney J.M.: The caloric equivalent of fever. II. Influence of major trauma. Ann. Surg. 1965; 161:140-147
9. Macklem P.T., Cohen C., Zagelbaum G., Roussos: The pathophysiology of inspiratory muscle fatigue. In: *Human Muscle Fatigue: Physiological Mechanisms*. London, Pitman Medical, 1981; p. 249
10. Harpin R.P., Baker J.P., Downer J.P., Whitwell J., Gallacher W.N.: Correlation of the oxygen cost of breathing and length of weaning from mechanical ventilation. Crit. Care Med. 1987; 15:807-812
11. Gilston A.: ARDS: Another approach? Intensive and Crit. Care Digest. 1985; 4:1-3
12. Moore F.D., et al.: *Post-Traumatic Pulmonary Insufficiency*. Philadelphia, WB Saunders Company, 1969
13. Rie M.A., Wilson R.S.: Acute respiratory failure. In: Tinker J., Rapin M. (Eds.) *Care of the Critically Ill Patient*. Berlin-New York, Springer-Verlag, 1983
14. Staub N.C.: Pathophysiology of acute lung injury. In: Vincent J.L. (Ed.) *Update in Intensive Care and Emergency Medicine*. Berlin-New York, Springer-Verlag, 1985; 3
15. Zapol W.M., Trelstad R.L., Snider M.T., Pontoppidan H., Lemaire F.: Pathophysiologic pathways

of the adult respiratory distress syndrome. In: Tinker J., Rapin M. (Eds.) *Care of the Critically Ill Patient*. Berlin-New York, Springer-Verlag, 1983; p. 341

16. Demling R.H., Flynn J.T.: Humoral factors and lung injury during shock, trauma and sepsis. In: Cowley R.A., Trump B.F. (Eds.) *Pathophysiology of Shock, Anoxia and Ischemia*. Baltimore-London, Williams & Wilkins, 1982; p. 395

17. Rinaldo J.E.: The prognosis of the adult respiratory distress syndrome. Inappropriate pessimism? Chest 1986; 90: 471-472

18. Cerra F.B., Border J.R., McMenamy R.H., Siegel J.H.: Multiple systems organ failure. In: Cowley R.A., Trump B.F. (Eds.) *Pathophysiology of Shock, Anoxia and Ischemia*. Baltimore-London, Williams & Wilkins, 1982; p. 254

19. National Heart, Lung and Blood Institute, Division of Lung Diseases: Extracorporeal support for respiratory insufficiency: A collaborative study. Bethesda, National Institutes of Health, 1979

20. Knaus W.A., Draaper E.A., Wagner D.P., Zimmerman J.E.: Prognosis in acute organ-system failure. Ann. Surg. 1985; 202:685-693

21. Goris R.J.A., te Boekhorst T.P.A., Nuytinck J.K.S., Gimbrere J.S.F.: Multiple-organ failure. Arch. Surg. 1985; 120:1109-1115

22. Askanazi J., Weissman C., Rosenbaum S.H., Hyman A.I., Milic-Emili J., Kinney J.M.: Nutrition and the respiratory system. Crit. Care. Med. 1982; 10:163-172

23. Kinney J.M., Weissman C., Askanazi J.: Influence of nutrients on ventilation. Reviews in Clin. Nutr. 1984; 54:917-929

24. Weissman C., Hyman A.I.: Nutritional care of the critically ill patient with respiratory failure. Crit. Care Clinics 1987; 3:185-203

25. Breadle D.L., Samsel R.W., Schumacker P.T., Cain S.M.: Critical O_2 delivery to skeletal muscle at high and low PO_2 in endotoxemic dogs. J. Appl. Physiol. 1989; 66 (6): 2553-2558

26. Weissman C., Forse R.A., Milic-Emili J., Askanazi J., Hyman A.I., Kinney J.M.: The metabolic and ventilatory response to the infusion of stress hormones. Ann. Surg. 1986; 203:408-412

27. Baue A.E.: The gut with injury: Problems and challenges. In: Baue A.E. (Ed.) *The Gastrointestinal Response to Injury, Starvation and Enteral Nutrition. Report of the eighth Ross Conference on Medical Research*. Columbus, OH, Ross Laboratories 1988;1

15. Metabolic Support and Energy Supply for Acute Respiratory Failure

G. Iapichino

ICU "E. Vecla", Institute of Anesthesiology and Intensive Care, University of Milan, IRCCS, Ospedale Maggiore, Milan, Italy

Malnutrition, pneumonia, sepsis, respiratory and multiple organ failures are the major complications and causes of death in critically ill patients.[1] Particularly, pneumonia and respiratory failure highly correlates with protein calorie malnutrition[2-4] that can impair respiratory muscle function, ventilatory drive and pulmonary defense mechanism.[5-7]

Interestingly enough most of the causes of the first two pulmonary alterations can be restored by nutritional repletion[1,8-12] even if it does not necessary follow that improving the nutritional status will improve the prognoses.[13-15]

Nevertheless many Authors believe that nutritional therapy should be given to patients who are or would be at risk of metabolic complication from starving.[5-7, 16-18] However, the feeding of patients with impending respiratory insufficiency or just on a ventilator is a complicated matter. We had in fact to consider the thermic effect of nutrients (DIT), gas exchange, substrate oxidation and direct effects on the lung, and finally the protein sparing effect of nutrition.

Thermic Effect of Nutrients

DIT i.e. the energy needed for absorption, processing and storage of nutrients that constitute the rate of energy expenditure (EE) increase above the premeal base line.[19]

In a normal volunteer DIT is about 10-15% of E intake with a meal;[20] about 10%[20] up to 20% after 5-7 days[21] during a continuous enteral infusion of a balanced diet at dosage 2-3 times resting EE (REE). On the contrary, infusion at a rate exactly balancing the REE seems to abolish DIT[20] in depleted stable patients. DIT depends also on the nutrients ranging from 30-40% for proteins to 6-8% carbohydrates and

2-3% lipids.[19] As a consequence, DIT is higher with high carbohydrate formula than with high fat formula either in depleted (32% vs 17% of E intake after 3 days of full enteral infusion[22]) or in injured patients (14-16% vs 10% of fasting REE during TPN)[18,23,24] (Fig. 1).

This finding means that we have the possibility of nutritionally modulating overall EE.

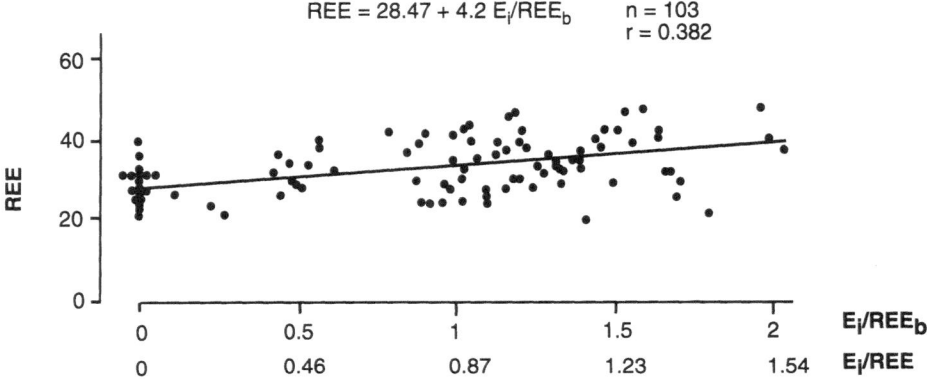

REE = 28.47 + 4.2 E$_i$/REE$_b$ n = 103   r = 0.382

Fig. 1 Total REE (kcal kg^{-1}/day) as a function of energy intake expressed as ratio to both basal fasting REE (Ei/REEb) and daily measured REE (Ei/REE) during TPN (glucose system) of catabolic patients. Reported by Iapichino et al.[39]

Gas Exchange

Increased EE obviously increases oxygen consumption ($\dot{V}O_2$) and carbon dioxide production ($\dot{V}CO_2$) i. e. cardiac output and ventilation, also promotes heat in suboptimal hydratation and circulatory conditions[25] (Fig. 2).

Gas exchange largely reflects the composition of the burned fuels; in comparison with carbohydrate, protein oxidation is associated with a 16% per kcal increase in $\dot{V}O_2$ while lipid with a 7.4% per kcal $\dot{V}O_2$ increase. On the contrary, $\dot{V}CO_2$ per available kcal decreases: 6,6% with protein and 24% with fat oxidation.[26]

Again, nutritional intervention allows modulation of cardiopulmonary workload.

Respiratory Quotient

The $\dot{V}CO_2/\dot{V}O_2$ ratio (R) is characteristic and constant for a given substrate oxidation, but R being only a ratio, cannot provide information about absolute gas exchange rate, the only variable of clinical relevance for cardiopulmonary function.[27]

Table I. Immediate oxidation of exogenous i.v. glucose determined by isotope studies as percentage of the intake.

Sepsis[34-36] 37-62%	burn[40] 56%	volunteer[3-4] 20%

Table II. Immediate oxidation of exogenous i.v. LCT emulsion determined by isotope studies as a percentage of the intake.

Age	Premature[35] 27% (4h)		Adult[28] 28% (7.5h)
Metabolic state	*Volunteer[28]* *28%*		*Trauma/Sepsis[28]* *37%*
Glucose intake mg kg^{-1} .min	0	2.5	5
depleted[28]	-	22%	9%
Trauma/Sepsis[28]	37%	26%	14%
Burn[29]			traces

Table III. Energy (kcal kg^{-1}) drown from oxidation of glucose and fat as calculated by isotope studies and Rnp.

	Expected REE	**Glucose input**	**EE from glucose (isotope)**	**(Rnp)**
Burn[40]	30	30	21	28 (94%)
Sepsis[34]	30	38	15	30 (100%)
Trauma[32]	28	15	9	17 (64%)
	Expected REE	**Fat input**	**EE from fat (C^{14} -palmitate)**	**(Rnp)**
Burn[29]	30	13	±	9 (29%)
Trauma[32]	28	15	10	10 (36%)

Moreover, R provides useful information together with the urinary nitrogen content, to determine the net proportion of carbohydrate, lipid and protein utilized to meet E requirements. Studies on R variations with different calorie substrates, suggest anyway a rank of preference in drawing E from different substrates. Reviewing this topic in catabolic patients[18] it appeared that exogenous fat availability does not affect glucose oxidation, i.e. adding lipid infusion to a given glucose load (less or equal to REE) does not modify R (Figs. 3,4,5). On the contrary, increased glucose availability reduces the fat utilization.[28,29]

This glucose preferential metabolism was isotopically confirmed in normal humans[30] in septic, cancer,[31] burn[29] and trauma[32] patients.

Besides, R does not indicate if the oxidized fuels become body stores or from the administered diet more lower isotope studies have shown that the amount of exogenous glucose or fat that is oxidized is only a part of the infused one[28,29,33-36] (Tables I, II). Moreover, oxidation derived from isotopes is always lower than the oxidation computed from calorimetry (Table III).

Rapidly available calorie substrates rank from long chain saturated, the worset[37], to glucose i.e. the same rank as for preferential E sources. Up to now, we need some more data about the degree and the meaning of medium chain saturated triglycerides oxidation[38] (Fig. 6).

Fuel storage and stored fuel oxidation is then a common effect of nutrition.

This finding means that we have also the possibility of choosing between an E

Table IV. Effect of i.v. glucose infusion on protein metabolism when the intake is in the range 0 to 4 and 4 to 8 mg kg^{-1} min^{-1}.

	Volunteers[36]		Depleted[39]		Injured[39]*				Septic[36]	
					M		S			
Intake	0-4	4-8	0-4	4-8	0-4	4-8	0-4	4-8	0-4	4-8
Ei/REE#			0.8	1.6	0.7	1.5	0.6	1.3		
Urea production%	-13	-28	-36	-23	-23	-18	-26	-19	-8	-31

* Moderately catabolic patients (M) had measured/predicted REE ratios greater than 1 but lower than 1.23; severely catabolic patients (S) had ratios greater than 1.23.

When reported, refers to the greater value of glucose intake.

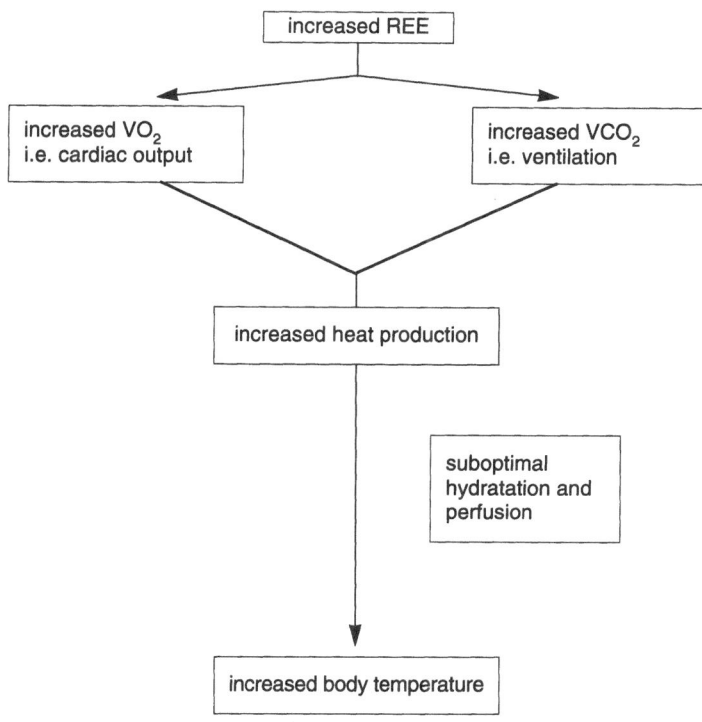

Fig. 2. Physiological consequences of increased REE.

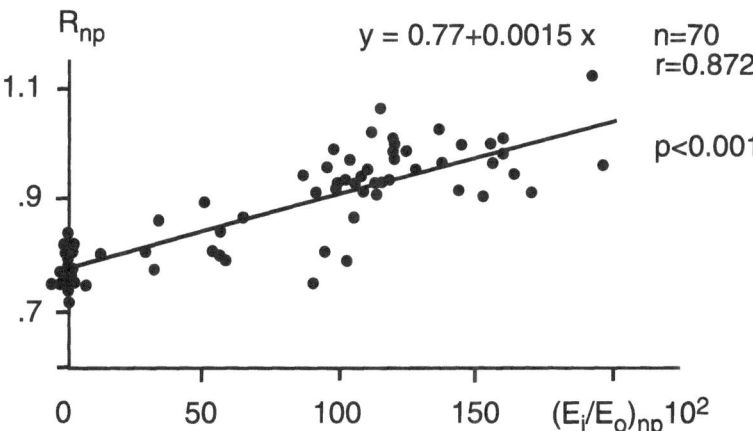

Fig 3. RQ as a function of energy intake (Ei) expressed as a percent of energy expenditure (Eo). The values refer to daily non protein (np) measurements during TPN (glucose system) of catabolic patients reported by Iapichino et al.[69]

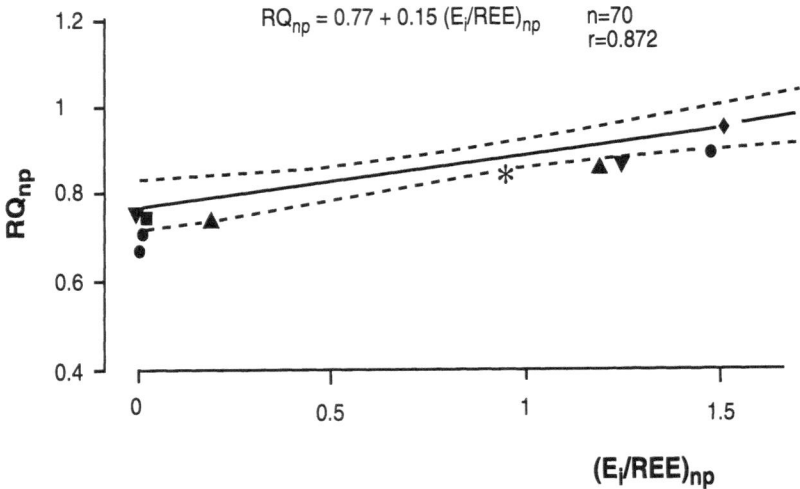

Fig. 4. Relationship and 99% confidence limits between RQ and energy intake (Ei) expressed as a ratio Ei to REE (Ei/REE). The values refer to daily non-protein (np) measurements during TPN (glucose system) of catabolic patients reported by Iapichino et al.[69] Closed symbols indicate literature calculated values from glucose system TPN. ▼[70], ●[46, 47], ▲[24], ■[29]. Data from ◆[71], ✳[72] refer to total RQ and Einp/REE.

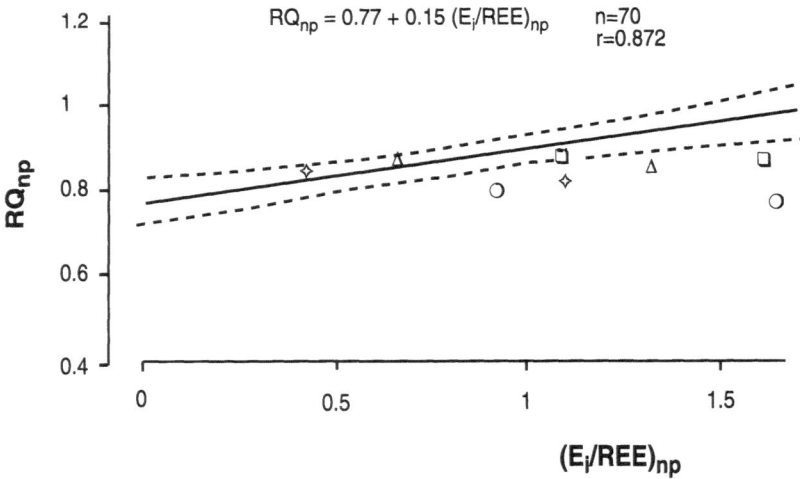

Fig. 5. Relationship and 99% confidence limits between RQ and energy intake (Ei) expressed as a ratio Ei to REE (Ei/REE). The values refer to daily non-protein (np) measurements during TPN (glucose system) of catabolic patients reported by Iapichino et al.[69] Open symbols indicate literature calculated values from mixed system TPN. Semi-open symbols indicate the RQ of mixed system TPN when the (Ei/REE) np is calculated from glucose intake only. ○[46, 47], Δ[24], ▢[29]. Data from ◇[73] refer to total RQ and Einp/REE.

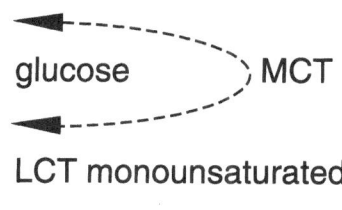

glucose   ⟩ MCT

LCT monounsaturated

LCT polyunsaturated

LCT full saturated

Fig. 6. Rank of prompt oxidability of exogenous non-protein substrates[28-38]

source mainly replacing the stores, even if probably oxidized after unknown day-time (fat), and a source with a relatively prompter availability (carbohydrate).

Protein Metabolism

Another effect to consider is the ability of the energy source to reduce protein losses and/or promote the retention of exogenous nitrogen, thus optimizing nitrogen utilization. The administration of i.v. glucose to healthy subjects[36] and to depleted[39] or catabolic patients[36,39] reduces nitrogen losses. The greatest sparing action is achieved with an energy load of between 1 and 1.3 times the REE (i.e. glucose 4,5 and 6 mg kg^{-1} min^{-1} in volunteers, depleted and catabolic patients respectively) (Table IV). Many Authors emphasize the specific effect of glucose with insulin.[36,40-44] The combined action could explain most of the papers reporting a significantly better protein sparing action of glucose than of mixed systems during TPN in catabolic patients.[41,43,45] Nordenstrom reported no statistical difference, only a trend in favour of glucose [46,47] and the only report[24] in which a mixed system was found superior, was obtained in a group of patients that was inhomogeneous for catabolism and nutritional status.

The results available from studies in depleted patients indicate the substantial equivalence of the two systems (for a review see[48]) despite the disagreement between important groups, for example, the difference in opinion between Shitzgal (better with glucose[49]) and McFie (better with a mixed system[50]). However two reports suggest that the insulin action may explain the possible difference between the two systems. The glucose system (better nitrogen balance) leads to higher insulin plasma levels than the mixed system.[48] With glucose the insulin levels are comparable with those which have been shown to be optimal for equilibrating protein synthesis and catabolism.[51]

During enteral feeding the energy source seems to play only a minor role in nitrogen metabolism, at least in depleted patients.[22]

Clinical Approach

Going into more detail of a well articulated approach to respiratory patients tailored on physiological premises but also on patient's conditions and different therapeutic goals, we can discuss some clinical situations. In the case of acute lung injury (ARDS) patients with a reduced healthy lung parenchyma for diffuse endothelial lesions with increased permeability (low compliance high shunt and dead space) coupled with an increased REE and protein catabolism, it would be mandatory not to increase $\dot{V}O_2$, $\dot{V}CO_2$, ventilation, ventilation/perfusion maldistribution and airway pressures. However, the satisfaction also of E need with rapidly available calories (Fig. 6) and the curtailing of protein loss is vital, being a matter of maintenance of physiological balance of such a critically ill patients. Then the suggestion could be 1000-1300 kcal, preferentially as glucose coupled with insulin (caution had to be also suggested in such a diseased lung for the safety[52-59] of LTC emulsion) and amino acid by vein. It could be also particularly useful to increase the nitrogen intake moderately (nitrogen 37 μg kg^{-1} .min has the same effect on nitrogen balance as glucose 4 mg kg^{-1} .min[60,61]) and to enhance the anabolic action of insulin with agents such as fructose disphosphate[62], ornithine-ketoglu-tarate[63] or glutamine.[64]

Nevertheless, the best nutritional support considering DIT, gas exchange and protein sparing effects, is to shift in a balanced enteral feeding as early as possible, trying to adequate calorie load to measured REE. We applied this technique with good gas exchange results also to patients with severe ARDS on extracorporeal CO_2 removal.[65]

In case of acute ventilatory failure in well nourished, eumetabolic patients we can agree with Koretz[15] and wait 5-7 days for the improvement of general and respiratory conditions before starting nutrition. If no improvement in ventilatory capacity is found, we have to remember that nutritional intake can increase ventilatory drive accelerating weaning from ventilator.[66,67]

Finally, depleted eumetabolic or moderately hypermetabolic patients had to be immediately treated with at least the needed E intake with the purpose of restoring as much as possible the metabolic derangements and the lean body mass for a successful weaning.[2]

We have the following theorical nutritional possibility:

- to supply E at a rate below or not exceeding REE trying to avoid net substrate storage thermic effect. This would minimize (TPN) or abolish (continuous enteral feeding) DIT (i.e. $\dot{V}O_2$ and $\dot{V}CO_2$), at least in unstressed people.
- to give the full calorie load (about 1,2-1,5 REE) as a high fat enteral formula diet.[68]

This approach minimized DIT of the selected calorie load, reduces $\dot{V}CO_2$, contrarywise $\dot{V}O_2$ will be increased and net protein utilisation will be not significantly worse with respect to a more balanced diet.[22] Exogenous lipid will mainly replenish body fat stores that are continuously mobilized to meet the EE that is not satisfied by the low glucose load. It is obvious that the patients must have enough fat store to immediately cope with daily E requirements.

The complete net replacement of protein and E stores with an extra needed fuel supply would be postponed until ventilatory failure is subsided.

References

1. Benotti P., Blackburn G.L.: Protein and or macro nutrient metabolic derangement of the critically ill patient. Crit. Care Med. 1979; 7:520-525
2. Kelly S.M., Rosa A., Field S., Coughlin M., Shizgal H.M., Macklem P.T.: Inspiratory muscle strength and body composition in patients receiving total parenteral nutrition therapy. Am. Rev. Respir. Dis. 1984; 130:33-37
3. Weissman C., Hyman A.I.: Nutritional care of the critically ill patient with respiratory failure. Crit. Care Clinics 1987; 185-203
4. Windsor J.A., Hill G.L.: Risk factors for postoperative pneumonia. Ann. Surg. 1988; 208:209-214
5. Askanazi J., Weissman C., Rosenbaum S.H., Hyman A.I., Milic-Emili J., Kinney J.M.: Nutrition and the respiratory system. Crit. Care Med. 1982; 10:163-172
6. Rochester D.F., Esau S.A.: Malnutrition and the respiratory system. Chest 1984; 85:411-415
7. Pingleton S.K., Harmon G.S.: Nutritional management in acute respiratory failure. JAMA 1987; 257:3094-3099
8. Doekel R.C., Zwillich C.W., Scoggin C.H., Krygger M., Weil J.V.: Clinical semistarvation: depression of hypoxic ventilatory response. N. Engl. J. Med. 1976; 295:358-361
9. Gertz I., Hedenstierna G., Hellers G., Wahren J.: Muscle metabolism in patients with chronic obstructive lung disease and acute respiratory failure. Clin. Sci. Mol. Med. 1977; 52:395-403
10. Weissman C., Askanazi J., Rosembaum S., Hyman A.I., Milic-Emili J., Kinney J.M.: Amino acids and respiration. Ann. Int. Med. 1983; 98:41-44
11. Aubier M., Murciano D., Lecocguic Y., Viires N., Jacquens Y., Squara P., Pariente R.: Effect of hypophosphatemia on diaphragmatic contractility in patients with acute respiratory failure. N. Engl. J. Med. 1985; 313:420-424
12. Wilson D.O., Rogers R.M., Sanders M.H., Pennock B.E., Reilly J.J.: Nutritional intervention in malnourished patients with emphysema. Am. Rev. Respir. Dis. 1986; 134:672-677
13. Koretz R.L.: Breathing and feeding. Can you have one without the other? Chest 1984; 86:298-299
14. Koretz R.L.: Nutritional support. Whether or not some is good, more is not better. Chest 1985;

88:2-3

15. Koretz R.L.: Breathing and feeding (Correspondence). Chest 1984, 86:651-652

16. Wilmore D.W.: Are the metabolic alterations associated with critical illness related to the hormonal environment? Clin. Nutr. 1986; 5:9-19

17. Cerra F.B.: Hypermetabolism, organ failure, and metabolic support. Surgery 1987; 101:1-14

18. Iapichino G., Radrizzani D.: Metabolic support and energy supply for critically ill patients: a pathophysiological approach. Intensive Care World 1988; 5:48-51

19. Jéquier E.: The influence of nutrient administration on energy expenditure in man. Clin Nutr. 1986; 5:181-186

20. Heymsfield S.B., Hill J.O., Evert M., Casper K., Digirolamo M.: Energy expenditure during continuous intragastric infusion of fuel. Am. J. Clin. Nutr. 1987; 45:526-533

21. Heymsfield S.B., Casper K., Funfar J.: Physiologic response and clinical implications of nutritional support. Am. J. Cardiol. 1987; 60:750-810

22. Heymsfield S.B., Head C.A., Mcmanus III C.B., Seitz S., Staton G.W., Grossman G.D.: Respiratory, cardiovascular, and metabolic effects of enteral hyperalimentation: influence of formula dose and composition. Am. J. Clin. Nutr. 1984; 40:116-130

23. Wilmore D.W.: *The metabolic management of the critically ill.* New York, Plenum Medical Book Co. 1977

24. Roulet M., Detsky A.S., Marliss E.B., Todd T.R.F., Mahon W.A., Anderson C.H., Stewart S., Feejeebhoy K.N.: A controlled trial of the effect of parenteral nutritional support on patients with respiratory failure and sepsis. Clin. Nutr. 1983; 2:97-105

25. Askanazi J., Rosenbaum S.H., Michelsen C.B., Elwyn D.H., Hyman A.I., Kinney J.M.: Increased body temperature secondary to total parenteral nutrition. Critical Care Med. 1980; 8:736-737

26. Diem K.: Documenta Geigy: *Tables scientifiques.* Basel, Geigy JR 1963; pp. 637

27. Silberman H., Silberman A.W.: Parenteral nutrition, biochemistry and respiratory gas exchange. JPEN 1986; 10:151-154

28. Nordenstrom J., Carpentier Y.A., Askanazi J., Robin A.P., Elwyn D.H., Hensle T.W., Kinney J.M.: Metabolic utilization of intravenous fat emulsion during total parenteral nutrition. Ann. Surg. 1982; 196:221-231

29. Goodenough R.D., Wolfe R.R.: Effect of total parenteral nutrition on free fatty acid metabolism in burned patients. JPEN 1984; 8:357-360

30. Wolfe B.M., Klein S., Peters E.J., Schmidt B.F., Wolfe R.R.: Effect of elevated free fatty acids on glucose oxidation in normal humans. Metabolism 1988; 37:323-329

31. Shaw J.H.F., Holdaway C.M.: Protein-sparing effect of substrate infusion in surgical patients is governed by the clinical state, and not by the individual substrate infused. JPEN 1988; 12:433-440

32. Shaw J.H.F., Wolfe R.R.: An integrated analysis of glucose, fat, and protein metabolism in severely traumatized patients: studies in the basal state and the response to total parenteral nutrition. Ann. Surg. 1989; 209:63-72

33. Eckart J., Tempel G., Kaul A., Witzke G., Schurbrand P., Schaaf H.: Metabolism of radioactive-labeled fat emulsions in the postoperative and posttraumatic period. Am. J. Clin. Nutr. 1973; 26:578-582

34. Wolfe R.R., O'Donnell T.F., Stone M.D., Richmand D.A., Burke J.F.: Investigation of factors determining the optimal glucose infusion rate in total parenteral nutrition. Metabolism 1980; 29:892-899

35. Paust H., Park W., Rating D., Helge H.: Measurement of fatty acid oxidation in premature newborn infants with the C-Triolein Breath Test. Clin. Nutr. 1984; 3:89-92

36. Shaw J.H.F., Klein S., Wolfe R.R.: Assessment of alanine, urea, and glucose interrelationships

in normal subjects and in patients with sepsis with stable isotopic tracers. Surgery 1985; 97:557-567

37. Jones P.J.H., Pencharz P.B., Clandinin M.T.: Whole body oxidation of dietary fatty acids: implications for energy utilization. Am. J. Clin. Nutr. 1985; 42:769-777

38. Bach C.A., Storck D., Meraihi Z.: Medium-chain triglyceride-based fat emulsions: an alternative energy supply in stress and sepsis. JPEN 1988; 12:82S-88S

39. Iapichino G., Gattinoni L., Solca M., Radrizzani D., Zucchetti M., Langer M., Vesconi S.: Protein sparing and protein replacement in acutely injured patients during TPN with and without amino acid supply. Intensive Care Med. 1982; 8:25-31

40. Burke J.F., Wolfe R.R., Mullany C.J., Matherws D.E., Bier D.M.: Glucose requirements following burn injury. Ann. Surg. 1979; 190:274-285

41. Long J.M., Wilmore D.W., Mason A.D., Pruitt B.A.Jr.: Effect of carbohydrate and fat intake on nitrogen excretion during total intravenous feeding. Ann. Surg. 1977; 185:417-422

42. Hinton P., Allison S.P., Littlejohn S., Lloyd J.: Insulin and glucose to reduce catabolic response to injury in burned patients. Lancet 1971; 1:767-769

43. Woolfson A.M., Heatley R.V., Allison S.P.: Insulin to inhibit protein catabolism after injury. N. Engl. J. Med. 1979; 300:14-17

44. Inculet R.I., Finley R.J., Duff J.H., Pace R., Rose C., Groves A.C., Woolf L.I.: Insulin decreases muscle protein loss after operative trauma in men. Surgery 1986; 99:752-758

45. Hennenberg S., Eklund H., Stjernstrom H., Hellsing K., Sjolin H., Wiklund L.: Post-operative substrate utilization and gas exchange using two different TPN-systems: glucose versus fat. Clinical Nutrition 1985; 4:235-242

46. Nordenstrom J.: Utilization of exogenous and endogenous lipids for energy production during parenteral nutrition. Acta Chir. Scand. 1982; 510 (suppl):1-199

47. Nordenstrom J., Askanazi J:; Elwin D.H., Martin P., Carpentier Y.A., Robin A.P., Kinney J.M.: Nitrogen balance during total parenteral nutrition: glucose vs fat. Ann. Surg. 1983; 27-33

48. Iapichino g., Radrizzani D., Leoni L., Osti G., Scherini A., Colombo A., Ronzoni G.: A comparison of mixed and glucose systems in the total parenteral nutrition of malnourished patients. Clinical Nutrition 1987; 6:1-4

49. Shitzgal H.M., Forse R.A.: Protein and caloric requirements with total parenteral nutrition. Ann. Surg. 1980; 192:562-568

50. McFie J., Yule A.G., Hill G.R.: Effect of added insulin on a body composition of gastroenterologic patients receiving intravenous nutrition. A controlled clinical trial. Gastroenterology 1981; 81:285-289

51. Burt M.E., Stein T.P., Schwade J.G., Brennan M.F.: Whole body protein metabolism in cancer-bearing patients. Cancer 1984; 53:1246-1252

52. Hageman J.R., Mcculloch K., Gora P., Olsen E.K., Pachman L., Hunt C.E.: Intralipid alterations in pulmonary prostaglandin metabolism and gas exchange. Crit. Care Med. 1983; 11: 794-798

53. Schmidt B.F., Allen R., Chandler C.: Effects of intravenous fat emulsion on pulmonary function in adult respiratory distress syndrome patients. Surgical Forum 1986; 37:84-86

54. Cerra F.B., Alden P.A., Negro F., Billiar T., Svingen B.A., Licari J., Johnson S.B., Holman R.T.: Sepsis and exogenous lipid modulation. JPEN 1988; 12:63S-68S

55. Spitzer J.J., Bagby G.J., Meszaros K., Lang C.H.: Alterations in lipid and carbohydrate metabolism in sepsis. JPEN 1988; 12:53S-58S

56. Skeie B., Askanazi J., Rothkopf M.M., Rosenbaum S., Kvetan V., Thomashow B.: Intravenous fat emulsions and lung function: a review. Crit. Care Med. 1988; 16:183-193

57. Venus B., Prager R., Patel C.B., Sandoval E., Sloan P., Smith R.A.: Cardiopulmonary effects of intralipid infusion in critically ill patients. Crit. Care Med. 1988; 16:587-590

58. Wolfe B.M., Suda S.A.: Invited comment: investigative obstacles in the study of the metabolism and toxicity of lipids. JPEN 1988; 12:59S-61S

59. Venus B., Smith R.A., Patel C., Sandoval E.: Hemodynamics and gas exchange alterations during Intralipid infusion in patients with adult respiratory distress syndrome. Chest 1989; 95:1278-1281

60. Iapichino G., Radrizzani D., Solca M., Pesenti A., Gattinoni L., Ferro A., Leoni L., Langer M., Vesconi S., Damia G.: The main determinants of nitrogen balance during total parenteral nutrition in critically ill injured patients. Intensive Care Med. 1984; 10:251-254

61. Radrizzani D., Iapichino G., Scherini A., Ferrero P., Doldi S., Solca M., Colombo A., Leoni L., Damia G.: Main nitrogen balance determinants in malnourished patients. Intensive Care Med. 1986; 12: 308-311

62. Radrizzani D., Iapichino G., Bonetti B., Colombo A., Leoni L., Elli R., Ronzoni G., Damia G.: Parenteral nutrition of injured patients. Surg. Gynecol. Obstet. 1986; 163: 37-41

63. Cynober L., Blond F., Lioret N., Coudrey-Lucas C., Saizy R., Gibondeau I.: Arterio-venous differences in amino acids, glucose, lactate, and fatty acids in burn patients: effects of ornithine alpha ketoglutarate. Clinical Nutrition 1986; 5:221-226

64. Stehle P., Zander J., Mertes N., Albers S., Puchstein C.H., Lawin P., Furst P.: Effect of parenteral glutamine peptide supplements on muscle glutamine loss and nitrogen balance after major surgery. Lancet 1989; 1:231-233

65. Iapichino G., Pesenti A., Radrizzani D., Solca M., Pelizzola A., Gattinoni L.: Nutritional support to long-term anesthetized and curarized patients under extracorporeal respiratory assist for terminal pulmonary failure. JPEN 1983; 7:50-54

66. Laaban J.P., Lemaire F., Baron J.F., Trunet P., Harf A., Bonnet J.L., Teisseire B.: Influence of caloric intake on the respiratory mode during mandatory minute volume ventilation. Chest 1985; 87:67-72

67. van den Berg B., Stam H., Hop W.C.F.: Effects of dietary protein content on weaning from the ventilator. Clinical Nutrition 1989; 8:207-212

68. Al-Saady N.M., Blackmore C.M., Bennet D.: High fat, low carbohydrate, enteral feeding lowers PaCO$_2$ and reduces the period of ventilation in artificially ventilated patients. Intensive Care Med. 1989; 15:290-295

69. Iapichino G., Solca M., Radrizzani D., Gattinoni L.: Fuel mixture burned during total parenteral nutrition in injured patients. I.R.C.S. 1980; 8:353-354

70. Askanazi J., Carpentier Y.A., Elwyn D.H., Nordenstrom J., Jeevanadam M., Rosembaum S.H., Gump F.E., Kinney J.M.: Influence of total parenteral nutrition on fuel utilization in injury and sepsis. Ann. Surg. 1980; 191:40-46

71. Long C.L., Blakemore W., Merrick H., Grecos G., Dennis R., Hall T.: Effect of carbohydrate intake on respiratory quotients and energy expenditure in septic patients. Fed. Proc. 1985; 44:1146

72. Giovannini I., Boldrini G., Castagneto M., Sganga G., Nanni G., Pittiruti M., Castiglioni G.: Respiratory quotients and patterns of substrate utilization in human sepsis and trauma. JPEN 1983; 197:27-33

73. Askanazi J., Nordenstrom J., Rosembaum S.E., Elwyn D.H., Hyman A.I., Carpentier Y.A., Kinney J.M.: Nutrition for the patient with respiratory failure: glucose vs fat. Anesthesiology 1981; 54:373-377

16. Cachexia in Chronic Obstructive Pulmonary Disease

E.F.W. WOUTERS, A.M.W.J. SCHOLS

Department of Pulmonary Diseases, University of Limburg, Maastricht, Netherlands

For many years, clinicians have recognized a gradual and significant weight loss and cachexia in a substantial number of patients with chronic obstructive pulmonary disease (COPD) during the natural course of their illness. Many studies have shown an increased incidence of cor pulmonale, heart failure and mortality in weight losing COPD patients.[1,4]

Many mechanisms have been postulated to explain the progressive weight loss in these patients: impaired gastro-intestinal function, a decreased dietary intake, an adaptive mechanism to decrease total oxygen consumption, a hypermetabolic state, depression and psychosocial factors.[5] They can be roughly divided in 2 groups: those regarding energy input and those regarding energy output.

Weight loss indicates an imbalance between energy input and energy output. In the present paper, commonly proposed theories of malnutrition and weight loss in patients with COPD will be discussed.

Energy Expenditure and Dietary Intake

The contribution of an increased energy expenditure and a decreased dietary intake to weight loss was studied in 80 elderly patients (65 ± 7 years) with a marked airflow obstruction (mean FEV_1: 35% pred).[6] Resting energy expenditure (REE) was measured by indirect calorimetry using a ventilated hood system. Habitual dietary intake (E I) of the period prior to admission was estimated with a diet history method.[7] Body fat free mass (FFM) was measured by the bio electrical impedance technique. FFM was calculated using the sex specific equations by Lukaski[8] based upon a linear relationship of fat free mass to body height^2/resistance.

Thirty nine patients (49%) suffered weight loss (> 7% of initial body weight) in the previous year. The average current body weight (BWt) in the weight losing group was 8 kg (12%) below the initial stable body weight. The ratio of 24 hours energy intake to 24 hours REE in the weight losing group amounted to 119% and was significantly lower than in the weight stable group (134%, p < 0.05). REE/BWt was significantly higher in the weight losing group (p < 0.001) whereas no significant difference in energy intake expressed per kg BWt was established between the weight losing and the weight stable patients (Table I). REE remained significantly higher when expressed per kg FFM (31.5 ± 4.5 vs 28.9 ± 3.4, p < 0.001).

A considerable number of patients (41/80; 51%) proved to exhibit an increased REE amounting to more than 24 kcal/kg/24h (Table II).[9] Mean values for FEV_1 and maximal inspiratory mouth pressure were significantly lower in the hypermetabolic patients than in the normometabolic patients (p < 0.001).

Dietary intake was significantly higher (p < 0.001) in the hypermetabolic

Table I. Energy balance in weight losing and in weight stable patients

Energy balance	Weight loss*		p-value
	Yes (n=39)	No (n=41)	
E I / BWt (kcal/kg/24h)	31.9±9.4	30.6±7.8	
REE / BWt (kcal/kg/24h)	26.4±4.0	22.7±2.9	<0.001

E I = Energy Intake; REE = Resting Energy Expenditure; * Mean±SD

Table II. Pulmonary function and energy intake in hypermetabolic and normometabolic patients

	Resting energy expenditure		p value
	>24 kcal/kg (n=41)	≤24 kcal/kg (n=39)	
FEV_1 (%)	29.0±12.0	41.0±18.0	<0.001
P_i max (kPa)	3.4± 1.4	4.8± 0.9	<0.001
EI/BWt (kcal/kg/24h)	35.5± 6.7	26.8± 7.9	<0.001

patients, indicating that the patients try to balance the increased energy requirements by increasing dietary intake (Table II).

The present data indicate that an increased REE, the major determinant of total energy expenditure, is a primary event in the initiation of cachexia in patients with COPD. Differences in body composition can only partly explain the observed differences in REE among the patients.

Previous studies have also demonstrated that the energy requirements of malnourished patients with emphysema are greater than would be predicted from published standards of calculated energy equivalents.[10] Braun et al.[11] have found a significant correlation between body weight and oxygen consumption/Kg BWt at rest in COPD patients and they concluded that an increased caloric utilization without adequate compensation in dietary intake is the reason for nutritional depletion in these patients. Goldstein et al.[12] have measured resting energy expenditure in a small group of 10 COPD patients: they concluded that malnourished patients with COPD have an elevated resting energy expenditure and that an increased diet induced thermogenesis may contribute to weight loss in these patients.

Fitting et al.[13] also described an increase in REE to 117 % of predicted basal metabolic rate and to 125 % of control group values, measured in ten patients with COPD.

Our results indicate that the increased metabolic rate may be caused by an increased oxygen consumption of the respiratory muscles secondary to an increased resistive load and impaired efficiency of the respiratory muscles.

The finding of an increased REE in patients with a history of weight loss is in contrast with the decrease in REE which normally occurs during starvation and weight loss in healthy individuals.[14]

The increase in REE/kg FFM is also in contradiction with the formulated hypothesis that weight loss is an adaptive mechanism to decrease oxygen consumption in severe COPD. This theory has sound physiological basis from studies of semistarvation in normal humans where both weight loss and decreased total oxygen consumption exceed the decrease in body mass, suggesting an adaptive process.[15] In conclusion, normal to even higher quantitative dietary intake can be expected in patients with COPD relative to normal standards. Dietary intake data have to be expressed relative to measured energy expenditure data and not in absolute terms.

Meal Related Oxygen Desaturation

The contribution of arterial oxygen desaturation during meals in COPD patients was studied in a group of 24 patients divided equally into 3 subgroups:

- group I: stable body weight and normal arterial oxygen tension (PaO_2 > 7.3 kPa).

- group II: recent involuntary weight loss but normal oxygen tension.
- group III: arterial oxygen tension less than 7,3 kPa.[16]

Oxygen saturation (SaO_2) was measured every minute by pulse oximetry before, during a standard meal and 30 minutes afterwards.

The average baseline saturation fell slightly during the meal in groups I and II, but this decrease was no different between the weight stable and weight losing patients.

On the contrary, most hypoxaemic patients exhibited a considerable desaturation of more than 5 percent. Contrary to the study by Brown[17], we did not find a significant correlation between baseline SaO_2 and the magnitude of desaturation. Arterial oxygen desaturation during meals may therefore only be a contributing factor to increase the imbalance between energy intake and energy expenditure in hypoxemic patients. The mechanism of this meal associated desaturation remains to be solved.

Other Factors

It has been suggested in the past that COPD patients were more prone to gastro-intestinal disorders than the general population.[18-20] None of the studies of gastro-intestinal disease and COPD had looked at the nutritional status and the role played by these types of gastro-intestinal symptoms in the weight loss of COPD remains to be clarified. However, the finding of a normal or even higher dietary intake in COPD patients compared to normal subjects suggests that these factors have only a limited importance in the weight loss of stable COPD patients. Psychosocial factors may also interfere with dietary intake. Braun et al.[11] have found that behavioural changes are of minor importance in the overall nutritional picture. In this study, only depression was directly related to body weight percent.

In conclusion, an increase in resting energy expenditure is a primary event in the development of malnutrition and weight loss in patients with COPD. The increased energy expenditure can be compensated by an increased energy intake.

The most likely cause of an increased energy expenditure in COPD is thought to be the respiratory muscles, facing a high load and hyperinflation. Meal related oxygen desaturation may contribute to impaired energy intake in hypoxemic patients.

References

1. Boushy S.F., Adhikair P.K., Sakamoto A., Lewis B.: Factors affecting prognosis in emphysema. Dis. Chest 1964; 45:402-11

2. Sukalmalchantra Y., Williams M.: Serial studies of pulmonary function in patients with chronic obstructive pulmonary disease. Am. J. Med. 1965; 39:941-45

3. Renzetti A.D., Mc Clement J.H., Litt B.D.: The Veterans Administration Cooperative Study of pulmonary function. Mortality in relation to respiratory function in chronic obstructive pulmonary disease. Am. J. Med. 1966; 41:115-129

4. Vandenbergh E., Van de Woestijne K., Gyselen A.: Weight changes in the terminal stages of chronic obstructive lung disease. Am. Rev. Resp. Dis. 1967; 95:556-566

5. Wilson D.O., Rogers R.M., Hoffman R.M.: Nutrition in chronic lung disease. Am. Rev. Resp. Dis. 1985; 132:1347-1365

6. Schols A.M.W.J., Mostert R., Soeters P.B., Greve L.H., Wouters E.F.M.: Resting energy expenditure in COPD. Europ. Resp. J. 1989; 2 (5)

7. Black G.: A review of validations of dietary assessment methods. Am. J. Epidemiol. 1982; 115:492-505

8. Lukaski H.C., Johnson P.E., Bolonchuk W.W., Lykken G.I.: Assessment of fat-free mass using bioelectrical impedance measurements of the human body. Am. J. Clin. Nutr. 1985; 41:810-817

9. Wahlquist H.L., Vobecky J.S.: Clinical nutrition problem evaluation and solving. In: Wahlquist H.L., Vobecky J.S., (Eds.) *Patient problems in clinical nutrition.* London, John Libbey & Company Ltd, 1988; 10-20

10. Wilson D.O., Rogers R.M., Sanders M.H., Pennock B.E., Reilly J.J.: Nutritional intervention in malnourished patients with emphysema. Am. Rev. Resp. Dis. 1986; 134:672-677

11. Braun S.R., Keim N.L., Dixon R.M., Clagnaz P., Anderegg A., Shrago E.S.: The prevalence and determinants of nutritional changes in chronic obstructive pulmonary disease. Chest 1984; 86 (4):558-563

12. Goldstein S., Askanazi J., Weissman C., Thomashow B., Kinney J.M.: Energy expenditure in patients with chronic obstructive pulmonary disease. Chest 1987; 91 (2):222-224

13. Fitting J.W., Frascarolo Ph, Jéquier E., Leuenberger Ph.: Energy expenditure and rib cage-abnominal motion in chronic obstructive pulmonary disease. Europ. Resp. J. 1989; 2:840-845

14. Brennan M.F.:Uncomplicated starvation versus cancer cachexia. Cancer 1977; 37:2359-2364

15. Keys A., Brozek J., Henschel A., et al.: *The biology of human starvation.* Minneapolis, The University of Minnesota. Press, 1950

16. Schols A.M.W.J., Cobben N., Mostert R., Wouters E.F.M.: Arterial oxygen saturation and carbon dioxide tension during meals in patients with severe chronic obstructive pulmonary disease. Chest 1991 in press.

17. Brown S.E., Casciatori R.J., Light R.W.: Arterial oxygen desaturation during meals in patients with severe chronic obstructive pulmonary disease. South Med. J. 1983; 76:194-198

18. Weber J.M., Gregg L.: The coincidence of benign gastric ulcer and chronic pulmonary disease. Ann. Intern. Med. 1955; 42:1026-1030

19. Zasly L., Baum G.I., Bumball J.M.: The incidence of peptic ulceration in chronic obstructive pulmonary emphysema. Dis. Chest 1960; 37:400-405

20. Browing R.J., Olsen A.M.: The functional gastrointestinal disorders of pulmonary emphysema. Mayo Clin. Proc. 1961; 36:537-543

Nutrition Therapy

17. Macronutrient Effect on Metabolism and Ventilation in Patients with Chronic Lung Disease

S. A. Goldstein-Shapses,[1] J. Askanazi[2]

1. Department of Orthopaedic Surgery, Columbia Presbyterian Medical Center, New York, USA
2. Division of Critical Care Medicine, Department of Anesthesiology, Montefiore Medical Center, New York, USA

Introduction

The nutritional complications in chronic obstructive pulmonary disease (COPD) often begin with weight loss, which occurs as the disease progresses. The weight loss not only results in malnutrition, but also contributes to deterioration of clinical and functional status, eventually leading to death.[1,2]

The aetiology of weight loss in patients with COPD has been attributed to:

1. inadequate nutrient intake
2. increased energy demands, as a result of inefficient use of the respiratory muscle.[3,4]

Most investigators agree that a raised metabolic rate (without compensation for energy intake) is largely responsible for the weight loss in patients with COPD.[4,8]

A sound understanding of the metabolism and ventilation in these patients before and after weight loss is the best method to approach their nutritional complications. Therefore, this chapter concentrates on the metabolic and ventilatory response to dietary intervention in patients with COPD and compares this to the response found in other disease states.

Metabolism

Metabolism refers to a wide range of biochemical interactions in the body. It would be impossible to analyze every detail of metabolism in a single patient. Therefore, metabolism is usually viewed on a whole body basis by measuring energy/nutrient intake and comparing it to energy output. Both energy intake and output can be varied by the investigator to obtain more information such as varying calorie and/or nutrient intake, or increasing stress, i.e. exercise, to vary energy output. Although many of the terms in this section of metabolism do not appear complicated, their definitions are often poorly understood. Consequently, some of these definitions are described below.

Depletion: refers to a reduction in body weight or lean body mass. Normally the depleted patient has lost 10% or more of their usual body weight. The depleted subject does not necessarily have to be under ideal weight, since an obese person that loses weight can become malnourished and remain overweight.

Resting Energy Expenditure (REE): energy expenditure during rest (after at least 30 minutes of lying quietly) at room temperature. It can be taken at any time of day and is independent of dietary intake. Basal energy expenditure (BEE or BMR) has more restrictions requiring that measurement be taken in the morning after an overnight fast.

Hypermetabolism: increased energy expenditure that is above the predicted values for a given energy intake.

Hypercatabolism: Increased N loss that is greater than normal for a given energy intake.

Nitrogen intake and utilization

In normal healthy subjects nitrogen (N) balance is regulated around zero, and is maintained within narrow limits. Fasting will result in negative N balance and net protein loss from the body, but the N excreted will diminish over time, conserving body protein.[9] Changes in N excretion will follow changes in N intake and vary with nutritional status. A high N intake in normal adults results in 7% retention of the N consumed,[10] whereas in depleted patients, retentions of up to 21% have been reported.[11]

Patients with COPD are hypermetabolic[4-7] with an energy expenditure approximately 15% above predicted (Harris-Benedict formula) or 26% above other depleted patients (Fig. 1). Since most hypermetabolic patients are also hypercatabolic (i.e. in sepsis or injury), it was surprising to find that our hypermetabolic patients with COPD had no evidence of excessive N excretion, and were therefore *not* hypercatabolic.[12] Depleted patients with emphysema can retain 15% of the N consumed on a high N and calorie intake, which is similar to depleted patients without emphysema receiving the same diet (Fig. 2).[12]

This indicates that hypermetabolic patients with emphysema are able to gain nitrogen at a similar rate to other malnourished patients who are hypometabolic. The results of dietary intervention can, therefore, be more promising for patients with COPD as compared to other hypermetabolic patients who can gain weight, but often continue to lose lean mass.

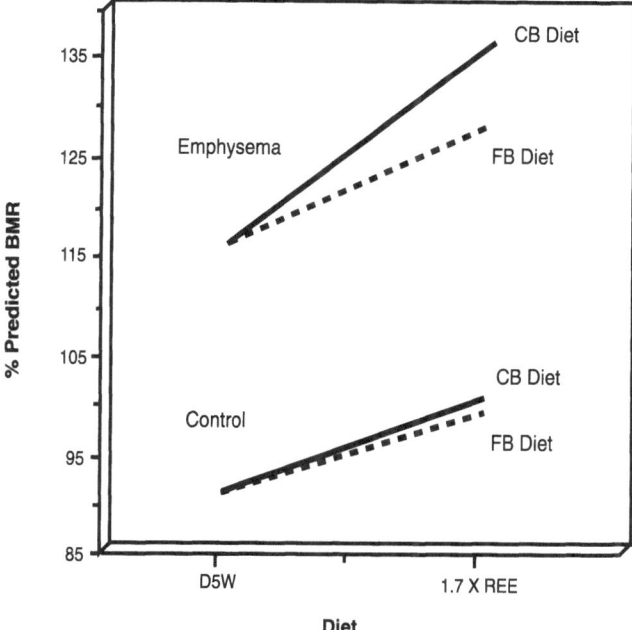

Fig. 1 Predicted energy expenditure in malnourished patients during 5% dextrose (D5W) and hypercaloric feeding with a carbohydrate based (CB) and fat based (FB) diet.

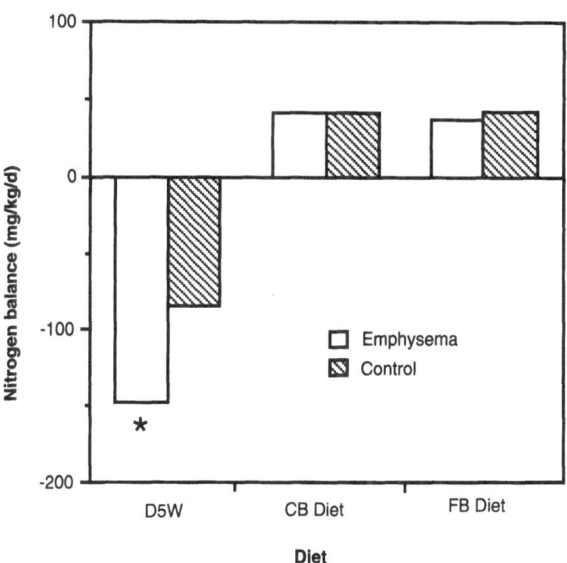

Fig. 2 Nitrogen balance in malnourished patients during 5% dextrose (D5W) and hypercaloric feeding with a carbohydrate based (CB) diet and fat based (FB) diet.
* - differs from control, p < 0.005

Energy intake and varying the non-protein energy sources

At a fixed energy intake, the level of energy intake and proportion of glucose and lipid will determine nitrogen retention. However, an increase in non-protein energy is not always as effective as protein supplementation, depending on the nutritional and metabolic state of the patient.

Fat and carbohydrate do not have equal potency in reversing the N loss due to starvation. At low energy intake, carbohydrate is more effective than fat in sparing protein. This is probably related to the minimum glucose requirements of the brain, which is 20% of REE. With a daily minimum glucose requirement of about 100-150 g/d, the N sparing efficiency of glucose and fat is similar in depleted patients.[13,15] Hypermetabolic septic and injured patients require more glucose (200+ g/d) to prevent catabolism.[16] Unlike these other hypermetabolic patients, depleted hypermetabolic patients with emphysema are able to retain N with a glucose intake as low as 100 g/d.[12] In this study[12], depleted patients with emphysema (who were not in any acute distress) were fed a hypercaloric (1.7 times REE) enteral or parenteral feeding regimen. These patients with emphysema retained a similar N balance when fed a carbohydrate based (53% carbohydrate, 30% fat and 17% protein) diet (Fig. 2). It is important to consider that the interactions between N and energy and the type of energy (glucose or fat) intake will determine the changes in body composition, and that this will not only vary between disease state, but also between individuals.

Ventilation

Nonprotein calories

Increasing glucose intake stimulates ventilation, secondary to an increased carbon dioxide production in normal subjects[17] and in patients with COPD.[18] The increase in ventilation is normally well tolerated by individuals without a limited respiratory reserve. However, in patients with lung disease, the increase in ventilation due to a high carbohydrate intake can result in serious and detrimental consequences to the patient.[19] In our experience of refeeding patients with emphysema, there remains a subset of these patients who are unable to tolerate a high caloric intake, i.e. greater than 1.5 times REE. These patients who cannot tolerate hypercaloric feeding often have symptoms of end-stage emphysema with evidence of cor pulmonale. Weight maintenance, rather than weight gain, should be the goal for depleted end-stage patients with emphysema.

Patients with COPD have a limited capacity for high level exercise after a bolus of glucose intake.[20] Depleted patients with emphysema who perform very low level exercise during hypercaloric feeding have an exaggerated ventilatory response to a carbohydrate based (CB) diet compared to a fat based diet.[18] In addition, hypercaloric feeding with a CB diet compared to a fat based diet appears to lower exercise efficiency in these patients. Intravenous lipids and high fat enteral formulas have been well tolerated by our patients with lung disease. This subject has been discussed in detail in another chapter.

Nitrogen intake

Amino acids will increase ventilation through an increase in ventilatory drive. An increased chemosensitivity to $PaCO_2$ appears to be responsible for the higher ventilation. Weissman et al. semi-starved normal subjects for one week on 5% dextrose and then infused an isocaloric solution of amino acids for 24 hours.[21] Within four hours, ventilation and oxygen consumption increased significantly, and increased further after 24 hours. The higher ventilation was primarily due to an increased inspiratory flow (tidal volume/time inspiration), and was associated with a decrease in exercise efficiency. Increasing the protein intake from 75 g to 150 g per day, while maintaining caloric intake will increase the ventilatory response to $PaCO_2$ in malnourished surgical patients.[22] This evidence suggests that protein intake increases ventilation through an increase in chemosensitivity, and appears to have a dose related response.

The macronutrient content of the diet has definitive effects on ventilation that may be beneficial or detrimental depending on the disease state. In patients with COPD, a high carbohydrate diet is probably detrimental by placing additional stress on an already overworked respiratory system. The extent of this stress (due to carbohydrate intake) has been demonstrated during maximal and low level exercise [18,20], and can be reduced with a fat based diet.[18]

Recommendations for feeding

Depleted patients with COPD should be encouraged to gain weight to prevent further lung deterioration and minimize the usual risks associated with malnutrition, such as infection and morbidity and mortality. An optimum caloric intake for malnourished patients with COPD is approximately 65-75% above their predicted REE, (or about 45 kcal/kg/d). A greater caloric intake can cause discomfort and dyspnea, whereas a lower intake may not promote weight gain or a positive N retention. Patients with end-stage emphysema usually have difficulty tolerating the ventilatory demands of a hypercaloric feeding regimen.

A caloric intake of 1.2 - 1.5 times predicted REE is probably the upper caloric limit in these end-stage patients.

When refeeding malnourished COPD patients with hypercaloric solutions, reducing the carbohydrate content will decrease the ventilatory requirements and may enhance efficiency during exertion, thereby promoting a more positive energy balance and weight gain. In light of this evidence, it appears that a fat based diet is advantageous over a carbohydrate based diet when refeeding malnourished patients with COPD.

172

Fat intake should be approximately 45-55% of the calories; any higher levels may be associated with reduced N retention or reduced fat clearance. Due to the increased risk factors associated with a high fat intake, this moderately high fat intake should be low in saturated fats, and is only recommended for the malnourished patient with COPD or in patients in acute respiratory distress. Nitrogen intake should be moderately high at approximately 300 mg/kg/d for patients with normal kidney function.

The long term effect (over the course of years) of weight gain in these patients on morbidity and mortality remains unresolved. Consequently, it remains unclear whether these patients benefit from weight gain or whether they can maintain the weight gain over an extended period of time.

Precise recommendations for nutritional therapy in all types of patients with COPD (i.e. malnourished, ventilator dependent, hypercapnic, etc.) will require an understanding of their individual metabolic and functional responses to feeding.

Future studies should focus on the metabolic responses to different feeding regimens in patients with varying degrees and types of lung disease. This could help to define the nutritional prescription for each patient type and the clinical benefits, if any, of weight gain in these patients.

References

1. Boushy S.F., Adhikair P.K., Sakamoto A., Lewis B.: Factors affecting prognosis in emphysema. Dis. Chest 1964; 45:402-411
2. Vandenbergh E., van de Woestijne K.P., Gyselen A.: Weight changes in the terminal stages of chronic obstructive pulmonary disease; relation to respiratory function and prognosis. Am. Rev. Resp. Dis. 1967; 95:556-66
3. Cherniack R.M.: The oxygen consumption and efficiency of the respiratory muscles in health and emphysema. J. Clin. Invest. 1959; 38:494-98
4. Donahoe M., Rogers R.M., Wison D.O., Pennock B.E.: Oxygen consumption of the respiratory muscles in normal and malnourished patients with chronic obstructive pulmonary disease. Am. Rev. Respir. Dis. 1989; 140:385-391
5. Goldstein S.A., Askanazi J., Weissman C., Thomashow B., Milic-Emili J., Kinney J.M.: Energy expenditure in patients with chronic obstructive pulmonary disease. Chest 1987; 91 (2):222-224
6. Hunter A.L.B., Carey M.A., Larsh H.W.: The nutritional status of patients with chronic obstructive pulmonary disease. Am. Rev. Resp. Dis. 1981; 124:376-381
7. Wilson D.O., Rogers R.M., Sanders M.H., Pennock B.E., Reilly J.J.: Nutritional intervention in malnourished patients with emphysema. Am. Rev. Respir. Dis. 1986; 134:672-677
8. Braun S.R., Keim N.L., Dixon R.M., Clagnez P., Anderegg A., Sharago E.S.: The prevalence and determinants of nutritional changes in chronic obstructive pulmonary disease. Chest 1984; 4:558-563
9. Benedict F.G., Carpentier T.M.: *A study of prolonged fasting*. Washington D.C., Carnegie Institute, Publ. N° 203, 1915
10. Oddoye E.A., Margen S.: Nitrogen balance studies in humans: long term effect of high N intake and N accretion. J. Nutr. 1979; 109:363-377

11. Shaw S., Elwyn D.H. Askanazi J., Kinney J.M., Iles M., Schwartz Y., Kinney J.M.: Effects of increasing nitrogen on N balance and energy expenditure in nutritionally depleted adult patients receiving parenteral nutrition. Am. J. Clin. Nutr. 1983; 37:930-940

12. Goldstein S.A., Thomashow B., Kvetan V., Askanazi J., Kinney J., Elwyn D.H.: Nitrogen and energy relationships and muscle strength when feeding malnourished patients with emphysema. Am. Rev. Respir. Dis. 1988; 138:636-644

13. Elwyn D.H., Gump F.E., Munro H.N., Iles M., Kinney J.M.: Changes in nitrogen balance of depleted patients with increasing infusions of glucose. Am. J. Nutr. 1976; 32:1597-1611

14. Nordenstrom J., Askanazi J., Elwyn D.H., Martin P., Carpentier Y.A., Robin A.P., Kinney J.M.: Nitrogen balance during total parenteral nutrition: Glucose vs. fat. Ann. Surg. 1983; 197 (1):27-33

15. Jeejeebhoy K.N., Anderson G.H., Nakhooda A.F., Greenberg G.R., Sanderson I., Marliss E.B.: Metabolic studies in total parenteral nutrition with lipid in man. Comparison with glucose. J. Clin. Invest. 1976; 57:125-136

16. Long C.L., Crosby F., Geiger J.W., Kinney J.M.: Parenteral nutrition in the septic patients: N balance, limiting plasma amino acids, and calorie to N ratios. Am. J. Nutr. 1976; 29:380-393

17. Angelillo V.A., Sukhdarshan B., Durfee D., Patterson A.J., O'Donohue W.J.: Effects of low and high carbohydrate feedings in ambulatory patients with chronic obstructive pulmonary disease and chronic hypercapnia. Ann. Int. Med. 1985; 103 (6 pt 1):883-885

18. Goldstein S.A., Thomashow B., Askanazi J., Milic-Emili J., Kinney J.M., Elwyn D.H.: Submaximal exercise in emphysema and malnutritional during two levels of carbohydrate and fat intake: J. Appl. Physiol. 1989; 67:1048-55

19. Askanazi J., Elwyn D.H., Silverberg P.A., Rosenbaum S.H., Kinney J.M.: Respiratory distress secondary to a high carbohydrate load: a case report. Surgery 1980; 87:596-598

20. Brown S.E., Wiener S., Brown R.A., Marcapelli P.A., Light R.W.: Exercise performance following a carbohydrate load in chronic airflow obstruction. J. Appl. Physiol. 1985; 58:1340-1346

21. Weissman C., Goldstein S.A., Askanazi J., Rosembaum S.H., Milic-Emili J., Kinney J.M.: Semistarvation and exercise. J. Appl. Physiol. 1986; 60 (6):2035-2039

22. Askanazi J., Weissman C., LaSala P., Milic-Emili J., Kinney J.M.: Effects of increasing protein intake on ventilatory drive. Anesthesiology 60:106-110

18. Parenteral Nutrition in Cystic Fibrosis

B. Skeie, E. Søreide, O. Kirvelä, D. P. Katz, J. Askanazi

Division of Critical Care Medicine, Department of Anesthesiology, Albert Einstein College of Medicine, Montefiore Medical Center, New York, USA

Patients with cystic fibrosis (CF) often present with malnutrition and pulmonary dysfunction. Pancreatic insufficiency with critical impairment of digestive function can be demonstrated in varying degrees, together with excess energy consumption due to the increased work of breathing and inadequate oral intake. Although pulmonary function has the greatest prognostic significance, Kraemer et al.[1] demonstrated a high correlation between nutritional status and survival in CF.

The mechanisms underlying the interaction between nutritional status and pulmonary function are manifold. Protein calorie malnutrition is believed to decrease the strength of respiratory muscles which results in diminished vital capacity, increased atelectasis and retained secretions, as well as impairment of the pulmonary parenchyma.[2,3]

Additionally, CF patients have abnormalities in fatty acid metabolism [4,6], including a relative deficiency in the essential fatty acids (EFA); this may cause an imbalance in eicosanoid (prostaglandins, leukotrienes, etc) production. Eicosanoids are important mediators in the inflammatory response and also affect bronchial and vascular muscle tone.[7] In chickens and rabbits, an EFA deficient diet can cause lung disease which has some similarity to CF.[8,9] Elliot reported that intravenous fatty acids in CF improved the fatty acid composition in plasma and the clinical course[10]; other trials have not shown uniformly positive results.[11,13] These studies differed in the length of treatment, the dose of fatty acids, the intervals between infusion, and the age and health of the patients.

It is clear that malnutrition is undesirable in patients with respiratory disease and significantly contributes to morbidity and mortality in the CF patient; several investigators have intervened with aggressive nutritional support. Shepherd et al.[14,15] studied 12 patients under the age of 12 for a period of 12 months, 6 months before and after they received a 21 day course of parenteral nutrition in addition to conventional therapy. The investigators found a favourable and persistent effect as measured by decreased number of respiratory infections, improved pulmonary function (FVC, FEV_1, PEF) and chest x-ray. Bertrand et al.[16] attempted to duplicate Shepherd's findings, using an enterally administered diet supplying a high caloric elemental intake provided at home for one month. By increasing the mean caloric intake to 133% REE, nutritional therapy resulted in a transient weight gain without any significant change in pulmonary function or clinical outcome.

Mansell et al.[17] studied the effects of parenteral nutrition in 11 patients in an older age group (10-17 years) for a one month period. In contrast to Shepherd's study, they found that a short course of parenteral nutrition can result in improved respiratory muscle strength assessed by improvement in maximum inspiratory pressure (MIP) and maximum breathing capacity (MBC), but no significant improvement in pulmonary function (assessed by VC, FEF, FEV_1 and FRC). However, these effects were transient and in most patients weight gain was lost within 6-12 months following parenteral therapy. The disparity in the results between this study and the more optimistic findings of Shepherd's group was felt to be related to age and rapidity in deterioration of pulmonary and nutritional status. It was suggested that aggressive nutritional support may be helpful in reversing an acute decline in pulmonary and nutritional status in younger patients but less helpful in reversing the insidious decline of the pulmonary and nutritional status of the older cystic fibrosis patient as studied in Mansell's group.

In the opinion of our nutrition group, long-term home parenteral nutrition (HPN) results in improvements beyond what can be explained by improved muscle function alone. HPN using a high percentage of the fat emulsion, Intralipid, appeared to result in clinical improvement in CF patients,[18,19] suggesting that the high concentration of linoleic acid in Intralipid could induce clinical benefit in CF via normalization of serum fatty acids (with concomitant effects on eicosanoid metabolism) in addition to improving the protein calorie status of the patients. This has been particularly dramatic in one patient carried for long term (2 years) on home TPN.[18] Slow infusion of lipid emulsion vasodilates the pulmonary circulation and may be therapeutic on that basis. The question of whether HPN with fat emulsion can have a more direct effect on pulmonary function, by providing precursors for synthesis of surfactant or prostaglandins, is also very important. We proposed that precise control of key infusion parameters (dose and type of fat, infusion rate, concomitant glucose concentration) appears to be important in obtaining a positive clinical effect.[7]

In a recent study we observed the effects of 4 months of HPN, consisting of Intralipid as the primary calorie source, in 20 CF patients.[20] The patients were randomly assigned to one of two groups. Group 1 started with HPN and group 2 was observed without HPN. After 4 months HPN was discontinued in group 1 and instituted in group 2. The questions we asked were:

1. Is there a benefit to the use of long term lipid based TPN to correct malnutrition across a broad population of patients with cystic fibrosis as assessed in terms of metabolic and functional parameters?
2. Is there a particular subgroup which will benefit maximally? Or conversely, are there some patients that should not have the therapy? Can these be identified prior to administration of therapy?
3. Do any observed beneficial effects indicate simply a benefit of nutritional support or is there evidence for a pharmacologic effect of lipid on lung inflammation?

Measurements were made of pulmonary function, exercise tolerance, metabolic and biochemical parameters at baseline, and at four months and eight months. This study was conducted in ambulatory patients at home. All the patients had Port-a-Caths or a similar central line placed at the start of the study so that home use of antibiotics was equivalent as far as could be determined.

We initially looked at the effects of the first four months of therapy and compared parenteral nutrition to the control. Not surprisingly the parenteral nutrition group showed a better response in terms of nutritional status than the control group. However, the response was seen mainly in triceps skinfold thickness (an indicator of fat deposition). Serum fatty acid profiles demonstrated some improvement in the group receiving parenteral nutrition; however, the control group also had some changes in a positive direction.

Pulmonary function tests demonstrated no significant changes in vital capacity or FEV_1 between the patients receiving parenteral nutrition or those in the control group. The other pulmonary function tests also showed no significant differences. The exercise test, in contrast to our previous anecdotal experience, deteriorated in the group receiving parenteral nutrition as it did in the control group.

Our conclusions for the first part of the study, in which we examined the effects of parenteral nutrition as compared to the control state, were that across this broad population of patients with cystic fibrosis, parenteral nutrition was really not very beneficial, given the small amount of clinical improvement that occurred along with the cost, discomfort and interference with patient sleep and so forth. Overall both the patients and the clinicians involved in the study felt the treatment was simply not worth the discomfort and possible complications associated with parenteral nutrition.

However, it was clear to us that there is a variability of response with some patients responding very dramatically in a positive manner and others having a very poor response. So, in our second analysis, we analysed the patients that received parenteral nutrition in an attempt to determinate whether there are biochemical parameters which can separate the "responders" from the "nonresponders". When we examined the group on the basis of serum fatty acid profiles it was apparent that those patients with a low but detectable concentration of gamma linolenic acid (DHLA) which returned to normal after 4 months, also responded in terms of physiologic and clinical functioning. Those in whom DHLA was not detectable at baseline or after 4 months of nutrition showed no improvement in clinical or physiologic function.

Vital capacity, FEV$_1$ and work function improved in the responders, but both work function and exercise time deteriorated in the nonresponders. The subjective responses of the patients agreed with the objective data. About 40% of the patients decided after a few months that the program resulted in an enormous benefit to them and improved their life quality; these patients elected to continue with the Intralipid based parenteral nutrition. About 60% told us that the program was not beneficial for them and elected to discontinue the parenteral nutrition after 4 months.

Our conclusions from this study are as follows:

1. There is a wide variability of responses to TPN in cystic fibrosis. Most patients did not improve their lung function after 4 months on an Intralipid based TPN therapy and we found no statistical significant differences between two groups in the pulmonary tests.
2. However, the variable response precludes our making broad statements at the present time regarding efficacy in the entire population.
3. Patients who request this therapy should be informed of the unpredictable nature of the response.

Our hypothesis for future research, based on this study, is that there is a defined metabolic defect that results in the failure of some patients with CF to respond to Intralipid-based parenteral nutrition. It would seem that a lowering of serum fatty acids occurs in CF that is detrimental to these patients.

In some patients, the lowering is primarily due to nutritional malabsorption phenomenon and these patients respond very nicely to intravenous replacement. There appears to be another group in whom, even when linoleic acid is normalized, there is an inability or decreased ability to convert linoleic acid to other fatty acids in the prostaglandin pathway. These patients respond poorly to nutritional intervention. We hypothesize that measurements of fatty acid metabolism would allow accurate prediction of which patients will respond poorly and are not candidates for parenteral nutrition with Intralipid. If the metabolic defect is the conversion of linoleic to gammalinolenic acid, the use of lipid emulsions containing either gammalinolenic or eicosapentaenoic acid may circumvent the metabolic defect and improve the response of patients who have a poor clinical result with Intralipid-based TPN.

References

1. Kraemer R., Rudebery A., Kadain B., et al.: Relative underweight in cystic fibrosis and its prognostic value. Acta Paediatric. Scand. 1978; 67:33
2. Rochester D.: Malnutrition and the respiratory muscles. In: Askanazi J (Ed) *Nutrition and respiratory disease*. Clinics in Chest Medicine 1986; 3:91-100
3. Sahebjami H.: Nutrition and pulmonary parenchyma. In: Askanazi J. (Ed) *Nutrition and respiratory disease*. Clinics in Chest Medicine 1986; 3:11-126
4. Kuo P.T., Huang N.N., Basset D.R.: The fatty acid composition of the serum chylomicrons and adipose tissue of children with cystic fibrosis of the pancreas. J. Pediatr. 1962; 60:394-403
5. Caren R., Corbo L.: Plasma fatty acids in pancreatic cystic fibrosis and liver disease. J. Clin. Endocrinol 1966; 26:470-7
6. Rosenlund M.L., Kim H.K., Kritchevsky D.: Essential fatty acids in cystic fibrosis. Nature 1974; 251:719
7. Skeie B., Askanazi J., Rothkopf M.M., et al.: Intravenous fat emulsions and lung function: a review. Crit. Care Med. 1988; 16:183-94
8. Craigh M.C., Faircloth S.A., Teer P.A., et al.: The essential fatty acid deficient chicken as a model for cystic fibrosis. Am. J. Clin. Nutr. 1986; 816-824
9. Harper T.B., Chase H.P., Henson et al.: Essential fatty acid deficiency in the rabbit as a model of nutritional impairment in cystic fibrosis. Am. Rev. Respir. Dis. 1982; 126:540-7
10. Elliot R.B.: A therapeutical trial of fatty acid supplementation in cystic fibrosis. Pediatrics 1976; 57:474-9
11. Davidson G.P., Phelan P.D., Townley R.R.W.: A controlled trial using intravenous infusion of soya oil emulsion in the treatment of children with cystic fibrosis. Austr. Pediatr. J. 1978; 64:207-13
12. Chase H.P., Cotton E.K., Elliot R.B.: Intravenous linoleic acid supplementation in children with cystic fibrosis. Pediatrics 1979; 64:207-213
13. Kusoffsky E., Strandvik B., Troell S.: Prospective study of fatty acid supplementation over 3 years in patients with cystic fibrosis. J. Ped. Gastroenter. Nutr. 1983; 2:434-8
14. Shepherd R.W., Holt T.L., Thomas B.J., et al.: Nutritional rehabilitation in cystic fibrosis: controlled studies of effects on nutritional growth retardation, body protein turnover, and course of pulmonary disease. J. Pediatr. 1986; 109:788-94
15. Shepherd R., Cooksley W.G.E., Cooke W.D.D.: Improved growth and clinical, nutritional and respiratory changes in response to nutritional therapy in cystic fibrosis. J. Ped. 1980; 97:351-7
16. Bertrand J.M., Marin C.L., Lasalle R., et al.: Short term clinical, nutritional and functional effects of continuous elemental, enteral alimentation in children with cystic fibrosis. J. Ped. 1984; 104:41
17. Mansell A.L., Andersen J.C., Muttart C.R., et al.: Short-term pulmonary effects of total parenteral nutrition in children with cystic fibrosis. J. Ped. 1984; 104:700-5
18. Askanazi J., Rothkopf M., Rosenbaum S.H., et al.: Treatment of cystic fibrosis with long-term home total parenteral nutrition. Nutrition 1987; 3:277-9
19. Skeie B., Askanazi J., Rothkopf M.M., et al.: Improved exercise tolerance with long-term parenteral nutrition in cystic fibrosis. Crit. Care Med. 1987; 15:960-2
20. Kirvelä O., Stern R.C., Askanazi J., et al.: Effects of long term parenteral nutrition in cystic fibrosis: A controlled clinical trial (In preparation).

19. Oral Intake, O_2, Dyspnea, Dysphagia and Other Considerations in Chronic Obstructive Pulmonary Disease Patients

R.D. FERRANTI,[1] M. ROBERTO,[2] C.BROWN III,[3] C.A.COELHO[4]

1. Center for Breathing Disorders, Gaylord Hospital, Yale University School of Medicine, USA
2. Department of Health Services State of Connecticut, USA
3. Pulmonary Section, Gaylord Hospital, State of Connecticut, USA
4. Communication Disorders Section, Gaylord Hospital, State of Connecticut, USA

Introduction

Confucius, the Chinese ethical teacher (551-479 B.C.) recommended "good food" and "breathing exercises" to gain good health.

Maimonedes (1135-1204) a renowned physician, philosopher and rabbi, was requested, while at a service of an Islamic patron, to prescribe a reliable dietetic regimen to relieve the attacks of shortness of breath, wheezing, and general weakness his employer often suffered after colds. Maimonedes wrote a treatise of 13 chapters; seven chapters related to diet and its significance in disease control. He stated that anything curable by food should not be treated otherwise.

In the last twenty years, two methods of treatment have gained increasing consideration in the long term care and rehabilitation of patients with chronic respiratory insufficiency:

- continuous low flow oxygen therapy;
- maintenance of good nutritional status and optimal weight by appropriate diet.

These approaches are easily understood. Oxygen and nutrients participate together and are necessary for the process of respiration. They are needed to furnish the energy necessary to perform the activities of daily living. Therefore, oxygen and nutrients need to be supplemented when they are insufficient to meet the demands of this work. In COPD patients with chronic respiratory insufficiency, this work may exceed the oxygen-nutrient availability.

Weight Loss and Visceral Protein Depletion in COPD

Several studies have demonstrated a correlation between worsening forced expiratory volume in one second, FEV_1, and percentage loss of body weight.[1,2,3] The *U.S. Veteran Administration Cooperative Study* on lung function found the following correlation.[4]

FEV_1	Average percentage of ideal body weight, (IBW)
> 1.49 L	97%
0.5 - 1.49 L	91%
< 0.5 L	82%

But even if, in COPD patients, there is a correlation between the deterioration in body weight, nutritional status and the increased work of breathing that the respiratory muscles need to generate as a consequence of the worsening obstructive condition, there is no evidence that other factors do not play an important role. Breathing pattern and nutritional status are not similarly affected in non COPD patients, such as in severe obesity, pregnancy and other conditions also with an increase in ventilatory load. Thus studies have been conducted to assess differences and find other factors which may affect breathing and nutrition in COPD patients.

A report of nutritional status in 107 COPD patients showed that 43% of these patients who had emphysema had less than 90% of IBW in contrast to only 3% of the patients with chronic bronchitis.[5]

During exercise gas exchange and dead space/tidal volume ratio worsen in the emphysema patient, but can improve in bronchitis and arterial oxygen tension (PaO_2) has a definite tendency to improve in bronchitis and worsen in emphysema (Fig. 1).[6] However it remains difficult, clinically, to clearly separate bronchitis from emphysema, and in the majority of the patients the two coexist.

It is also generally accepted that personality traits may influence physiological function. Patients with bronchitis have been classified as "blue bloaters" or non-fighters and patients with emphysema as "pink puffers" or fighters. Type A, or pink puffer, tend to try to push performance beyond their physiological capability. In cardiovascular diseases a performance/arousal curve has been suggested, in which metabolism reaches a point where it changes from anabolic to catabolic.[7]

Two studies investigated correlations between anthropometric measures, visceral proteins and lung function.

One study in 60 COPD outpatients suggested that body weight and anthropometric changes are related to oxygen consumption, VO_2 per kilogram (VO_2/kg), probably secondary to increase ventilatory need at a given VO_2; while biochemical parameters, in particular albumin levels appeared more closely related to airway

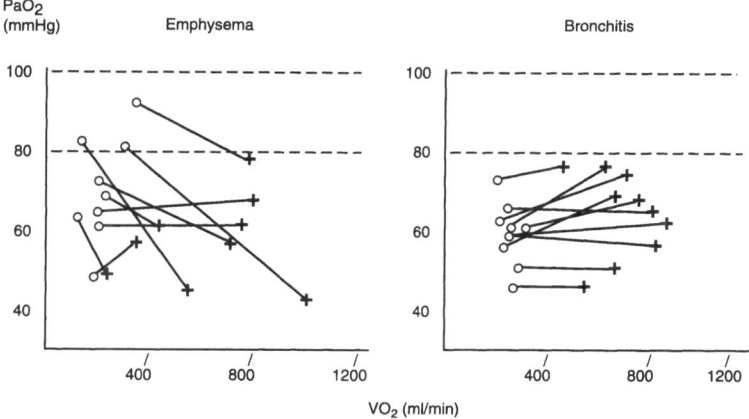

Fig. 1 Arterial PO$_2$ at rest (O) and during exercise (+) related to O$_2$ uptake in the two groups of patients. Dotted boundaries indicate normal range (+2SD from mean). (Modified by N.L. Jones)[6].

function and lung diffusion. This study also considered psychological factors such as depression and anxiety, which were found to play a lesser role in the nutritional status of the COPD patient.[8] In a second study of 153 COPD patients, the correlation between biochemical parameters and impaired gas exchange also showed to be more significant than with spirometric data, and increased protein depletion correlated better with hypoxia than with the degree of bronchial obstruction. In these patients with severe disease, in addition to a marasmic state there can be a kwarshiorkor state, and albumin appears a more discriminative indicator of visceral protein depletion, than prealbumin and transferrin.[9]

Albumin as an Indicator of Performance and Rehabilitation Potential in COPD Patients

An initial serum albumin value less than 2.5 gm/100ml was found not to be a good prognostic sign for survival[10], and the ability to respond to nutritional support with an increase in protein synthesis, was found to be a good prognostic sign for weaning from assisted ventilation after respiratory failure.[11]

A study was conducted in our rehabilitation hospital to examine the statistical relation between significant variables and outcomes of the inpatient COPD population. Our patients usually have a PaO$_2$< 55mmHg and are placed on oxygen supplement before, or immediately on admission. For the patients in the study the median FEV$_1$ was .560 L/sec and the VC median was 1410 ml. The serum albumin level median was 3.2 (Gm/1000ml) (range 1.7-6.5). In this study, albumin was proxy measure for nutritional status and proved to be a stronger predictor of outcome than FEV$_1$ or vital capacity. Initial serum albumin levels were recorded in

282 patients or 71% of the study population. The albumin levels correlated by discharge category, i.e., whether the patient was discharged to his/her own home, to skilled nursing facility, or had to be readmitted to the acute hospital for complications, or expired (Fig. 2) (p < .005). Transfer ability, a functional measure of ability to move from bed to chair, also showed a significant correlation with the admission serum albumin value. (R2 = .115; p < .0001).[12]

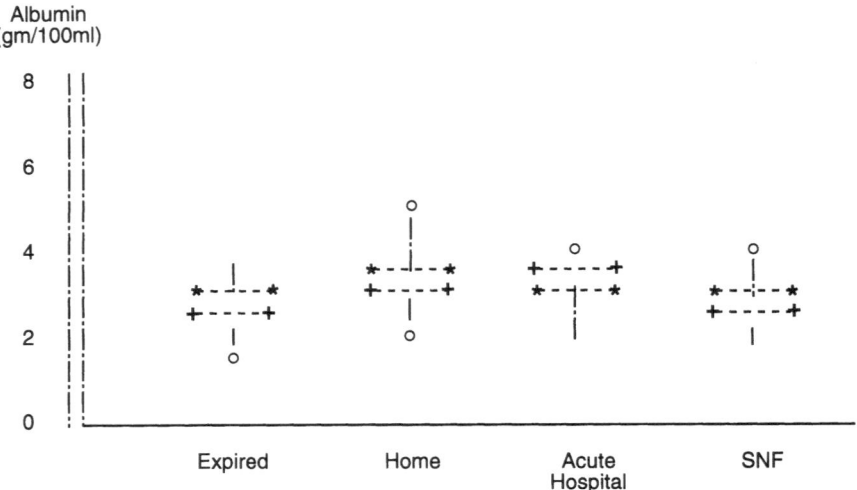

Fig. 2 Observed Albumin levels in the four different discharged categories.
Acute Hospital = Transfer due to complication; SNF = Skilled Nursing Facility for chronic care

Oxygen Desaturation, Dyspnea, Respiratory Rate and Anorexia

Hypoxaemia affects all organ systems. Neuropathies have been described in COPD patients with severe hypoxaemia.[13] According to the severity and duration of hypoxaemia, liver function can be affected in varied degrees.[14] Hypoxia in skeletal muscle may result in impaired mitocondrial oxidative function and low values for ATP and phosphocreatinine, and COPD patients have been found to demonstrate, both in their respiratory and non-respiratory muscles, reduction in ATP and phosphocreatinine values, without relation to the type and level of activity or the severity of COPD.[15]

The beneficial effect of oxygen supplementation in COPD patients was first reported in 1968; this included weight improvement.[16]

After beginning continuous oxygen therapy one can observe some improvement in the caloric consumption of several, but not all, underweight COPD patients.

However, until recently, oxygen supplement has been, and still is largely based on obtaining arterial blood gases at rest.

With the availability of oximetry trending it has been found that many COPD patients on oxygen supplement at a steady flow, may desaturate for significant parts of the day; during sleep, exercise and eating. In particular, desaturation during meals can be demonstrated in COPD patients.[17]

After reports that the use of a pendant oxymizer cannula protects COPD patients from desaturation during exercise,[18] we studied patients with poor caloric intake during meals, as measured by significant food left over in their trays, and when dyspnea appeared to interfere with their eating. These patients showed desaturation during eating, which also was amenable to improvement by the use of the pendant oxymizer; the improvement, on average, was of 6% up to 94% SaO_2 (p < .001) (Fig. 3).[19]

While further controlled studies should be conducted, prompt improvement in caloric intake also appeared evident. The mechanism for this may be that in improving oxygenation and reducing respiratory rate, competition between breathing and swallowing is reduced.

Breathing and swallowing are reciprocal acts, in order not to aspirate food in the

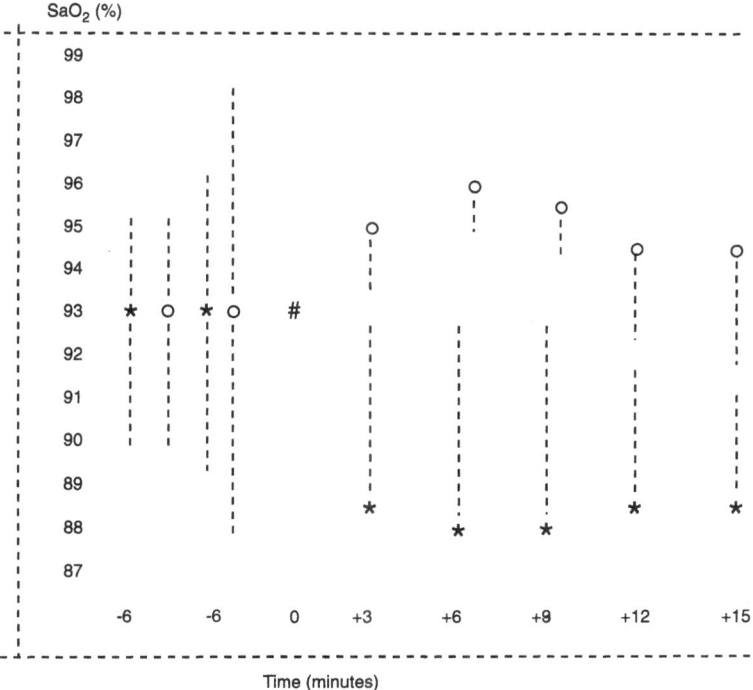

Fig. 3. Prevention of oxygen desaturation by oxymizer during food consumption. O Reservoir Cannula; * Regular Cannula; # Onset of food consumption

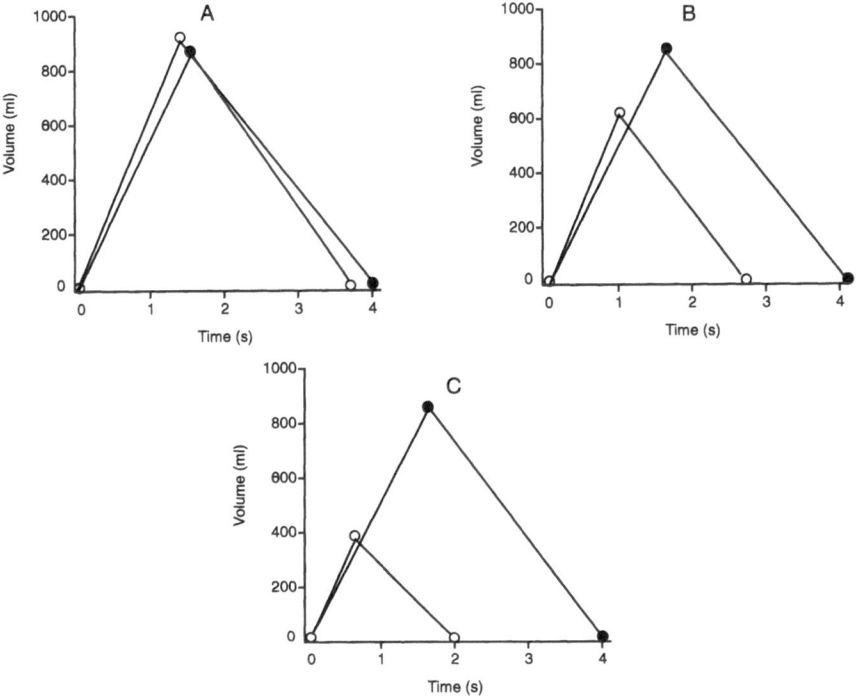

Fig. 4. Average respiratory cycle in patients with COPD (○), as compared to normal subjects (●) (modified by Milic-Emili [21]; A = non-hypercapnic COPD patients; B = hypercapnic COPD patients; C = COPD patients in acute respiratory failure

lungs one cannot breath and swallow at the same time.

The process of breathing can be better understood analyzing minute ventilation with respect to the duration and the rate at which air is inspired and expired.[20] The breathing pattern in COPD patients shows a prolonged exhalation at the expense of a reduced inhalation. Respiratory insufficiency, failure and hypercapnia enhance this pattern (Fig. 4).[21] Oxygen intake is decreased, and oxygen waste increased due to increasing dead space and expiratory time. In COPD patients on oxygen supplement on a steady flow rate, if the respiratory rate doubles, the flow rate must double to maintain the same oxygen tension and saturation in the blood.[22]

Oxymizer systems deliver a high oxygen pulse very early in inspiration, protect the patient against desaturation and appear to decrease competition between breathing and eating.

Other Causes of Anorexia, and Difficulty in Swallowing

Other causes of anorexia, which may be observed in COPD patients, are also common to other conditions.

Table I: Swallowing dysfunction observed in COPD patients.

		Dysfunction					
Subject	Aspiration	Bolus control	Oral transit	Swallow response	Pharyngeal transit	Laryngeal closure	Crico-pharyngeal
1		+	+		+		
2			+	+	+		
3			+	+	+		
4							
5	+	+	+		+		
6	+	+	+	+	+		
7			+		+		
8		+	+		+		
9							
10		+	+	+	+		
11		+	+	+	+		
12							
13	+		+		+		+
14							
	3	6	10	5	10	0	1
Total	21%	43%	71%	36%	43%		7%

- Active infections
- Side effects of medications
- Altered taste
- Digestive enzymes insufficiency and malabsorption
- Psychosocial factors
- Organic central nerves pathology
- Reflex avoidance secondary to diarrhea, others...

In patients with advanced respiratory insufficiency, especially after decompensation, intubation and/or tracheostomy and assisted ventilation, poor progression of the bolus through the pharynx can be observed. At our hospital we studied the swallowing mechanism of 14 patients who had been transferred to our centre because of a difficult and protracted recovery after respiratory failure.

Each patient received a bedside evaluation to assess the oral phase of the swallow, as well as video fluoroscopy during which patients were presented with three consistencies of contrast material to swallow: liquid barium, barium paste, and crackers coated with barium to examine the pharyngeal phase. Results indicated that nearly all of the patients experienced some difficulty with both phases of swallowing. Oral and pharyngeal transit times were consistently slower than normal. Most patients demonstrated diminished coordination and strength of the oral and pharyngeal musculature. The overall picture was one of reduced strength in all aspects of the swallow, coupled with a reduced ability to use pulmonary air to clear the larynx and ensure airway protection. Consistent aspiration was observed in only 3 of the 14 patients, but 10 of the patients were judged to have a moderate swallowing dysfunction (Table I).[23]

Conclusion

The relationship between breathing and swallowing needs to be closely investigated in patients with severe respiratory insufficiency.

In patients with COPD and nutritional depletion, especially after decompensation and respiratory failure it is important to optimize oxygenation, assess proper feeding and rule out dysphagia. Hospitalization, enteral and parenteral feeding may be necessary to re-establish nutritional compensation and assess the best possible maintenance rehabilitative regimen.

References

1. Bushi S.F., Adikari P.K., Sakamoto A., Lewis B.M.: Factors affecting prognosis in emphysema. Dis. Chest 1964; 45: 402-11
2. Sukumalchantra Y., Williams M.H.: Serial studies of pulmonary function in patients with chronic obstructive pulmonary disease. Am. J. Med. 1965; 39:941-945
3. Aieganspack D., Sunderman A., Rath W.: Kopergenichtbeim Lungemphysem (eine Klinisch - Statistiche Studie) Alerg. Asthma (Leipzig) 1963; 9,233 (Abstract)
4. Renzetti Jr.A.D., McClement J.H., Litt B.D.: The Veterans Administration Cooperative Study of pulmonary function. Am. J. Med 1966; 41: 115-129
5. Wilson D.O., Rogers R.M., Openbrier D.: Nutritional aspects of patients with chronic obstructive pulmonary disease. In: B.J. Make (Ed.) *Pulmonary Rehabilitation.* Clinics in Chest Med., 1986; 7:643-656
6. Jones N.L.: Pulmonary gas exchange during exercise in patients with chronic airway obstruction. Clin. Sci. 1966; 31:39-50
7. Nixon P.G.F.: The human function curve. Practitioner 1976; 217 (1301): 765-70, 217 (1302) 935-44
8. Braun S.R., Keim N.L., Dixon R.M., Anderegg A., Sgrage E.: The prevalence and determinants of nutritional changes in chronic obstructive pulmonary disease, Chest 1984; 86:558-563
9. Schols A., Mostert R., Soeters P. et al.: Inventory of Nutritional Status in patient with COPD. Chest 1989; 96:247-49
10. Apelgren R.N., Rombeau J.L., Twomey P.L., Miller R.A.: Comparison of nutritional indices and outcome in critically ill patients. Critical Care Medicine 1982; 10,55: 305-307

11. Larca L., Greenbaum D.M.: Effectiveness of intensive nutritional regimes in patients who fail to wean from mechanical ventilation. Critical Care Medicine, 1982; 10, 5: 297-300

12. Roberto M.: Predictors of outcome for COPD patients in a Rehabilitation Facility. Unpublished, Yale University 1989

13. Narayan M., Ferranti R.: Nerve conduction impairment in patients with respiratory insufficiency and severe hypoxemia. Arch. Phys. Med. Rehab. 1978; 59:188-192

14. Refsum H.E.: Severe Arterial Hypoxemia and Liver Cell Necrosis in patients with Pulmonary Insufficiency. Acta Med. Scand. 1964; 176, 4:473-478

15. Fiaccadori E., Del Canale S., Guariglia A.: Nutritional and metabolic aspects of COPD, In: Grassino A. et al. *Current Topics in Rehabilitation. Respiratory Muscles in Chronic Obstructive Pulmonary Disease*, London, Springer-Verlag Publishers 1988; 111-123

16. Petty T.L., Finnegan M.M.: Clinical evaluation of prolonged ambulatory Oxygen Therapy in Chronic Patients with Airway Obstruction. Am. J. Med. 1968; 45:242-252

17. Brown S.E., Casciari R.J., Light R.W.: Arterial oxygen saturation during meals in patients with severe chronic obstructive pulmonary disease. South Med. J. 1983; 76 (2):194-8

18. Arlati S., Rolo J., Micaleff E., Sacerdoti C., Brambilla I.: A Reservoir Nasal Cannula Improves Protection Given by Oxygen During Muscular Exercise in COPD, Chest 1988; 93:1165-1169

19. Brown C., Ferranti R.D., Sparapani M.: Reservoir nasal cannula prevents desaturation in COPD patients during eating. Am. Rev. Resp. Dis. 1988; 137, 4:157

20. Brancroft J., Margaria R.: Some effect of carbonic acid on the character of human respiration. J. Physiol. (London) 1931; 72:175-85

21. Milic-Emili J.: Recent advances in clinical assessment of control of breathing. Lung 1982; 160:1-17

22. Shigeoka J.W., Bonekat H.W.: The Current Status of Oxygen Conserving Devices, Respiratory Care 1985; 30:833-836

23. Coelho C.: Preliminary Findings on the nature of dysphagia in patients with chronic obstructive pulmonary disease. Dysphagia 1987; 2:28-31

20. Influence of Body Mass Index on Maximal Static Respiratory Pressures

N. D'ALOYA, S. SUBIACO, S. ANTENORI, L. LENTINI

Department of Respiratory Physiopathology, "A. Murri" Hospital, Jesi, Ancona, Italy

Introduction

Maximal static expiratory and inspiratory pressures (PE_{max} and PI_{max}) are regarded as simple and practical measurements which can help to evaluate respiratory muscle strength. Although affected by subject effort and despite wide inter- individual variability and wide between-study differences, the clinical use of these measurements is well established especially in monitoring the effectiveness of respiratory rehabilitation and in detecting respiratory muscle fatigue.[1]

Most of the available studies on maximal respiratory pressures take into consideration their relationship to age and sex[3], to lung function[5], to skeletal muscle strength[9] or to pathological situations.[2,4,7,8] Thus, a better knowledge about the influence of other anthropometric factors may lead to a more appropriate interpretation of these measurements.

The purpose of this study was to evaluate the influence of Body Mass Index (BMI) on PE_{max} and PI_{max}, in order to contribute to a more satisfactory definition of normal reference values.

Subjects and Methods

Two hundred and twenty six healthy subjects (163 males and 63 females, 18 to 69 years of age) underwent standard spirometry and P_{max} measurement, using the obstruction technique described by Black and Hyatt.[2]

Selection of healthy subjects was obtained by exclusion from the sample of those with abnormal chest findings (physical and X-ray examination) or with spirometric abnormalities (Vital Capacity/predicted, % : <80; Tiffeneau Index : <95).

Subjects were simply instructed to perform forced expirations against the obstructed airway starting from maximal lung filling level and forced inspirations from maximal emptying level. An air leakage through a 20-gauge needle was preventing facial muscles from producing added positive pressure. Distally to the mouthpiece and proximally to the obstruction, the airway was connected to two separate aneroid gauges for positive and negative pressure, equipped with maximum value indicating needle and manual zero resetting. Both devices underwent previously a semistatic electromechanical calibration (Hartmann & Braun /Shoppe & Faeser plunger movement pressure transducer TD/G 131-Oe) and the outcoming instrumental error was set within plus-minus 3% in both measuring ranges (0 to +300 and 0 to -300 cmH$_2$O).

P_{max} values found in the sample were slightly or moderately lower than Black and Hyatt's reference values for age and sex, as shown in the graphic comparison of figure 1. However, there was a substantial agreement in the feature of regressions.

By classifying the sample according to the value of BMI (Table I), only 115 subjects (about 51%) were included within the normality range in respect of this ponderal index.

Table I. Mean values and SD of Age and P_{max} found in 226 healthy adults, classified in 4 groups according to the value of BMI.

	BMI (Kg/sqm)	number	Age (year)	PE$_{max}$ (cmH$_2$O)	PI$_{max}$ (cmH$_2$O)
underweight	<19	8	23.9±5.6	126.3±34.6	81.9±20.0
normal weight	19-24	115	26.0±6.8	172.0±48.7	105.9±34.1
overweight	24-29	83	35.2±13.6	179.2±48.7	115.5±34.8
pathol. obese	>29	20	39.5±12.3	188.0±47.5	121.0±33.3

Thus, to simplify statistical analysis, the sample has been further divided just into two groups by a cut-off value of BMI=24, in order to allow a comparison between subjects with BMI<24 (n=123) and BMI>24 (n=103) as shown in tables II and III.

Results and Discussion

Evidence that BMI is an important determinant of P_{max} is suggested by the higher

Table II. Mean values and SD of anthropometric and lung function measurements in 226 healthy adults divided into 2 groups (cut-off value: BMI=24).

		Age (year)	BSA (smq)	BMI (Kg/smq)	VC (litre)	PE max (cmH_2O)	Pi max (cmH_2O)
A. BMI<24	x	25.8	1.72	21.77	4.89	169.1	104.3
(n=123)	SD	6.8	0.17	1.54	1.00	49.2	33.9
B. BMI>24	x	36.0	1.87	27.03	4.82	180.9	116.6
(n=103)	SD	13.5	0.15	2.95	1.07	48.6	34.6
						t = 1.808	2.686
						p : NS	<0.01

Table III. Comparison of both P_{max} regressions vs. Age (year), Body Surface Area (BSA, sqm), Vital Capacity (VC, litre) and Body Mass Index (BMI, Kg/sqm) in the two groups A) and B) of Table II.

	r	t	p	score
A. BMI<24 (n=123)				
PE_{max} vs. Age	0.099	1.104	0.3	NS
PI_{max} vs. Age	0.174	1.949	0.1	NS
PE_{max} vs. BSA	0.510	6.514	0.001	****
PI_{max} vs. BSA	0.431	5.248	0.001	****
PE_{max} vs. VC	0.526	6.800	0.001	****
PI_{max} vs. VC	0.467	5.811	0.001	****
PE_{max} vs. BMI	0.345	4.049	0.001	****
PI_{max} vs. BMI	0.297	6.523	0.001	****
B. BMI>24 (n=103)				
PE_{max} vs. Age	0.219	2.258	0.05	*
PI_{max} vs. Age	0.264	2.749	0.01	***
PE_{max} vs. BSA	0.403	4.430	0.001	****
PI_{max} vs. BSA	0.363	3.918	0.001	****
PE_{max} vs. VC	0.467	5.305	0.001	****
PI_{max} vs. VC	0.447	5.027	0.001	****
PEmax vs. BMI	0.099	0.999	0.4	NS
PImax vs. BMI	0.043	0.429	0.7	NS

score 0.05=*; 0.02=**; 0.01=***; 0.001=****

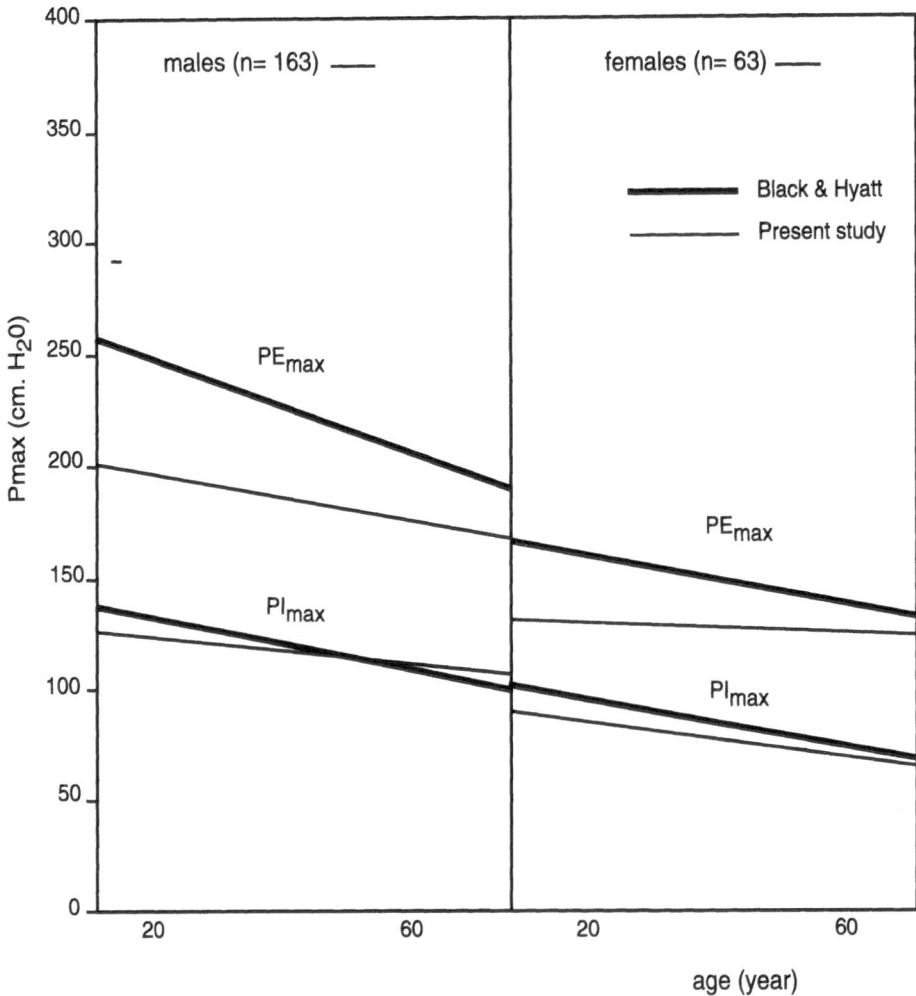

Fig. 1. Comparison with Black & Hyatt's equations.

values of both expiratory and inspiratory pressures in subjects with BMI>24 (Table II), although with significant differences between means only for PI_{max} (p<0.01). Thus, it is worth noting that PI_{max} might be likely to be more suitable than PE_{max} in detecting the influence of BMI on respiratory muscles strength.

Furthermore, the higher correlation coefficients of both P_{max} regressions versus BSA, VC and BMI compared with the insignificant ones of the regression versus age suggest (at least in the examined sample) that body size related indexes might

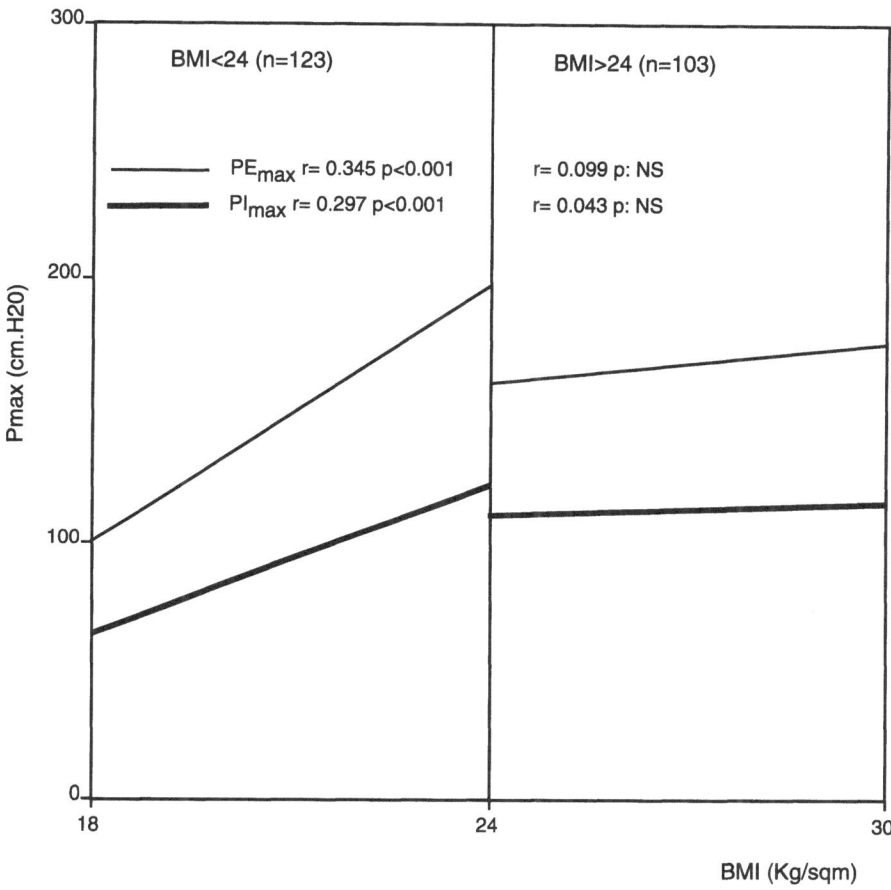

Fig. 2. PE_{max} and PI_{max} regressions vs. BMI (n=226)

play a more important role than age, among the determinants of P_{max} (Table III). Lack of correlation between P_{max} and BMI in subjects with BMI>24 (emphasized in Fig. 2) could be explained by mechanical disadvantage due to obesity.[6]

Conclusion

These findings suggest that BMI and other ponderal indexes might heavily affect P_{max} measurements. Thus, much more information concerning the influence of body size related anthropometric indexes seems to be needed for defining normal reference values of maximal respiratory pressures.

196

References

1. Ambrosino N., Nava S., Zocchi L.: La fatica muscolare respiratoria. Rass. Patol. Appar. Respir. 1989; 4: 110-115
2. Bird R.B., Hyatt R.E.: Maximal respiratory pressures in chronic obstructive lung disease. Am. Rev. Respir. Dis. 1968; 98: 848-856
3. Black L.F., Hyatt R.E.: Maximal respiratory pressures: normal values and relationship to age and sex. Am. Rev. Respir. Dis. 1969; 99: 696-70
4. Black L.F., Hyatt R.E.: Maximal static respiratory pressures in generalized neuromuscular disease. Am. Rev. Respir. Dis. 1971; 103: 641-650
5. Leech J.A., Ghezzo H., Stevens D., Becklake M.R.: Respiratory pressures and function in young adults. Am. Rev. Respir. Dis.1983; 128: 17-23
6. Melzer E., Souhrada J.F.: Decrease of respiratory muscle strength and static lung volumes in obese asthmatics. Am. Rev. Respir. Dis. 1980; 121: 17-22
7. Rochester D.F., Braun N.M.T., Arora N.S.: Respiratory muscle strength in chronic obstructive pulmonary disease. Am. Rev. Respir. Dis. 1979; 119 (Suppl: 151-154)
8. Szeinberg A., Mindorff C., Bryan A.C., England S., Levinson H.: Maximal inspiratory and expiratory pressures in cystic fibrosis. Am. Rev. Respir. Dis. 1984; 129 (Suppl. 207-ib)
9. Wagener J.S., Hibbert M.E., Landau L.I.: Maximal respiratory pressures in children. Am. Rev. Respir. Dis. 1984; 129: 873-875

21. The Effect of Resistive Breathing Training Combined with Nutritional Supplementation in Malnourished Chronic Obstructive Pulmonary Disease

D. Gross,[1] E. Meyron,[2] D. Schindler[3]

1. Department of Anesthesiology, Hadassah University Hospital, Jerusalem, Israel
2. Department of Dietary Services, Hadassah University Hospital, Jerusalem, Israel
3. Maccabi Sick Fund, Tel-Aviv, Israel

Introduction

Weight loss in some patients with Chronic Obstructive Pulmonary Disease (COPD) has been observed for many years. Malnutrition is commonly, but not inevitably associated with emphysemateous COPD.[1]

Significant weight loss has been reported in 27-71 percent of all patients.[2-4]

Patients who lose weight have increased incidence of cor-pulmonale and greater mortality.[4-8] The average life expectancy in the patients who begin to lose weight has been reported to be three years or less.[6]

This prognosis might be due to impairment of respiratory muscle structure and function in humans which in turn induces respiratory insufficiency.[9,10] Thus, the impact of malnutrition on the clinical course of a number of acute and chronic respiratory diseases maybe substantial. Hyperinflation as it occurs in COPD exerts a mechanical disadvantage condition on the respiratory muscle performance by a number of mechanisms.[11] Severe hyperinflation shortens and flattens the diaphragm, altering the length of its fibers and thus their force production, inducing an elevation of the work of breathing and energy expenditure.[12] It has also been shown that maximum inspiratory and expiratory forces correlate significantly with percentage of ideal body weight.[13]

Respiratory muscle strength has been associated with maximum voluntary ventilation (MVV), an index of ventilatory endurance.[10] In addition, nutritional supplementation to malnourished COPD has been demonstrated to improve inspiratory mouth and transdiaphragmatic pressure, indicating the improvement in ventilatory strength.[14]

The ventilatory muscles can be trained for both strength and endurance in normal subjects,[15] quadriplegic patients,[16] patients with cystic fibrosis[17] and patients with COPD.[18,19]

Skeletal muscle function depends on its functional energy stored and supplied and its mechanical efficiency,[20] namely, if the energy supplied to the muscles is reduced due to decreased blood flow, malnutrition or hypoxia, the muscle function will be impaired.[20] The respiratory muscles behave similarly.[21]

The purpose of the present investigation was to study the effect of resistive breathing training with nutritional supplementation on the respiratory muscle function and ventilatory capacity in malnourished COPD.

Method and Procedures

Preliminary investigations of five poorly nourished COPD patients were studied. The patients, physical characteristics and nutritional status are presented in Table I.

Table I. Physical and nutritional characteristics.

number	5
age	58.0± 7.3
weight (kg)	57.2± 2.5
height (cm)	176.4± 3.4
CHI (%)	71.6± 7.9
weight % ideal body weight	97.0± 7.1
AMC % standard	85.0±11.6

All subjects have not been smoking for the last two years or longer. The clinical and nutritional status of these patients was recorded on the first visit at the clinic. They were all severely obstructed and/or restricted as well. Their initial pulmonary functions are presented in Table II.

Nutritional Assessment

Nutritional measurements: height and body weight were measured and ideal body weight was estimated from sex and height according to a standard equation.[22]

Anthropometric measurements included the measured variables, triceps skinfold (TSF), and midarm circumference (MAC), and the derived variable, arm muscle circumference (AMC). All measurements were taken on the dominant arm at the

Table II. Baseline pulmonary functions

Subjects	FVC L. (%)		FEV L. (%)		MVV L. (%)	
1	2.23	(73)	2.20	(85)	68	(70)
2	1.02	(30)	0.42	(17)	19	(19)
3	1.60	(58)	1.50	(53)	32	(27)
4	1.50	(74)	1.03	(61)	38	(37)
5	1.90	(50)	0.83	(27)	36	(31)

midpoint between the acromion and alecranon process.[23] Triceps skinfold was measured using large skin calipers by pinching a fold of skin from the underlying triceps muscles. Midarm circumference was recorded by a soft tape measure calibrated in centimetres. It was placed firmly around the arm at its midpoint in the same place as previously described for skinfold.

Arm muscle circumference (AMC) was then calculated from values derived from TSF and MAC, as shown in the following equation, where all dimensions are in centimetres. All anthropometric measurements were performed by one person.

$$AMC = MAC\ (\pi \times TSF)$$

Creatinine height index (CHI) was computed from the following measurements: twenty four hour urine collection, from which a urine sample was obtained for urine analysis of creatinine content. The CHI was calculated based on the following equation:

$$CHI = \frac{100 \times \text{actual urinary creatinine (in 24 hours)}}{\text{ideal urinary creatinine (in 24 hours)}}$$

The CHI is an index of body muscle mass (derived from urinary creatinine).

The ideal urinary was obtained from a standard. Blood was observed for SMAC and albumin.

Ventilatory and Spirometric Function Tests

Patients had a spirometric evaluation, using the Vitalograph (compact model), on the first visiting day at the clinic. The spirometric test included slow and forced vital capacity (FVC), forced expiratory volume in one second (FEV$_1$), ratio of FEV$_1$/FVC, and forced expiratory flow between 25 and 75% of vital capacity.

An index of respiratory muscle endurance was estimated from the 12-s MVV (l/

min). Although MVV can be reduced either by abnormal lung mechanics or by the decrease in respiratory muscle strength, the decrease in MVV can be attributed to respiratory muscle weakness and reduced ventilatory endurance.[10]

The spirometric and MVV values were corrected to body temperature and pressure saturated with water vapor, and compared with their predicted values on the basis of age, sex and height equations.[24]

Procedures

After a complete evaluation patients received supplemental enriched diet for a period of two months. The diet was adapted individually to each patient. Revaluation was then performed and the treatment of resistive breathing was added to the supplemental diet. The resistive breathing was applied as previously recommended.[16] This combined treatment of resistive breathing and enriched diet continued for a period of two months, after which a complete revaluation was performed.

Results

Pulmonary function and nutritional data of the patients are presented in Table I and II. Patients at the beginning of the study had a restrictive (FVC<70%) pattern; Three of these patients had FVC of less than 60% of predicted values. All patients had also severe obstructive airway disease with FEV_1 less than 1.0 l. (less than 61% predicted)(Table II). Measurement of Maximum Voluntary Ventilation (MVV) was significantly lower than that of normal subjects (mean = 36.8±8.8% predicted)(Table II).

Patients were found to have mild to moderate malnutrition, with body weight % ideal body weight = 95.6±5%, values of skinfold were <80% of standard according to sex. Creatinine height index (CHI) was demonstrated to be <70%.

The CHI was significantly correlated with both FVC (r=.781, p<0.05) and MVV (r=.775, p<0.05) (Fig. 1). No correlation was found between any other nutritional parameter and the pulmonary functions. Nevertheless, the small increase of the nutritional parameters as a result of treatment was not significant, while the pulmonary functions indicating the development of respiratory muscle strength and endurance did indeed increase. The first two months of supplemental diet induced an increase in both FVC from 58.6 to 63.6 percent predicted (p<0.05), and MVV from 36.8 to 44.8 percent predicted (p<.01) (Fig. 2).

The combined treatment of resistive breathing training increased the FVC further to a value of 69.4 percent predicted (p<.02). This change was similar to that of the first two months of treatment (Fig. 2). The MVV after a combined treatment of resistive breathing and nutritional supplementation however increased to a value

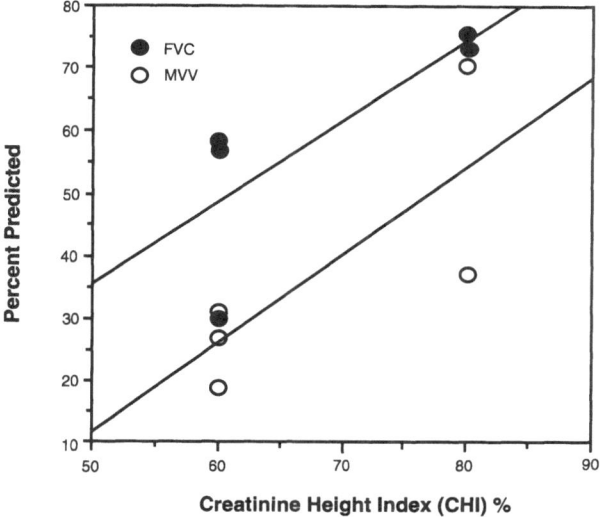

Fig. 1 Relationship between creatinine height index (CHI) and percent predicted values of Forced Vital Capacity (FVC) (●) and Maximum Voluntary Ventilation (MVV) (O).

of 57.8 percent predicted (p<.005). This increase was significantly greater than that observed during the first two months of treatment (nutritional supplementation only) (Fig. 2).

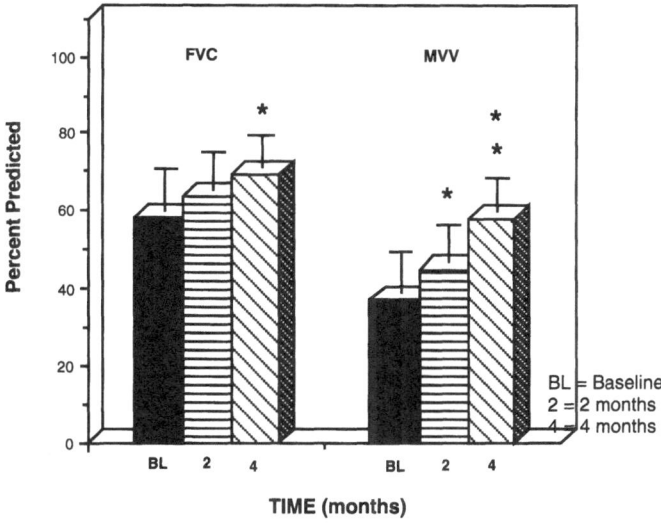

Fig. 2 The change in FVC and MVV after two months of nutritional supplementation[2] and after two months of combined treatment of nutritional supplementation with resistive breathing.[4] * = P<0.05, ** = P<0.01.

Discussion

The major findings of this study of ambulatory malnourished COPD patients were that:

1. All patients had severe COPD.
2. They had perhaps a slightly reduced weight, which did not correlate with the pulmonary functions and/or the ventilatory capacity.
3. The only nutritional parameter significantly correlated with the ventilatory capacity and pulmonary function was the Creatinine Height Index (CHI).
4. The treatment of nutritional supplementation (NS) improved the values of functional residual capacity (FVC), and the maximum voluntary ventilation (MVV).
5. The combined treatment of NS and resistive breathing (RB) was found to affect the FVC in the same way as the treatment during the first two months. However, the increase in MVV (and index of ventilatory muscle endurance) was greater as a result of the combined treatment (NS and RB) during the second two months than during the first two months.

We found no relationship between the nutritional parameters BMI, AMC, and skinfold and the pulmonary functions nor the ventilatory capacity and endurance. These findings are the same as previously described.[22] Nevertheless, we found good correlation between CHI which is an index of body muscle mass and the pulmonary functions, ventilatory capacity and endurance. Respiratory muscle strength has been previously shown to be significantly correlated with MVV (Fig. 3) hence, the high correlation between CHI and MVV would indicate a good relationship between respiratory muscle mass and ventilatory strength and endurance. The reduction in muscle mass is also indicated by a decrease in AMC to values lower than 85%.

Malnutrition is common among COPD patients. In our study these patients had reduced CHI, AMC and FFM as well as a very substantial reduction in FVC and MVV. These results indicate a reduction in muscle strength and endurance, influencing the patients performance[19] similarly to that reported by Arora and Rochester[10] and further confirmed by Wouters et al.[25]

Ventilatory muscle function depends on their functional energy stored and supplied, as well as their mechanical efficiency.[20] If the energy stored and supplied to the muscles is reduced due to decreased blood flow, malnutrition or hypoxia, the muscle function will be impaired.[20] Nutritional supplementation to malnourished COPD was shown to increase inspiratory mouth and transdiaphragmatic pressures, indicating an improvement in ventilatory strength[14] perhaps due to an increase in the potential energy supply to the periphery. Our results demonstrated a significant

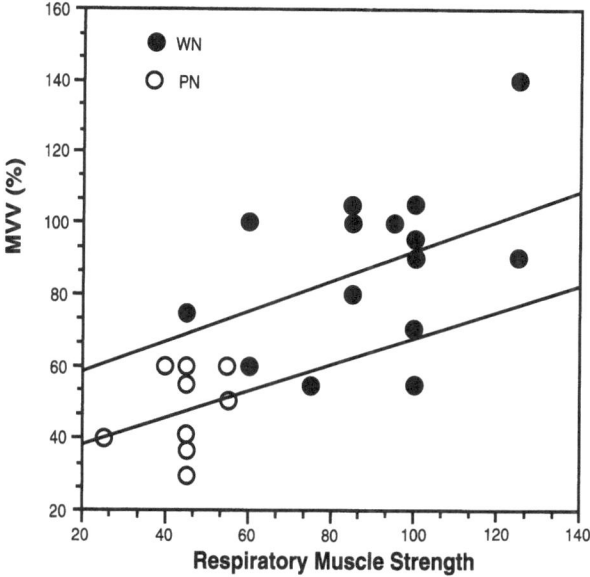

Fig. 3 The relationship between respiratory muscle strength and MVV percent predicted for the poorly nourished (PN) and well nourished (WN) COPD patients.

improvement in FVC and MVV with nutritional supplementation for a duration of two months, also depicting an increase in ventilatory muscle strength and endurance. Respiratory muscle strength is significantly correlated with MVV in both well and poorly nourished patients with COPD (Fig. 3). In addition, MVV had been shown too be the best parameter correlated with ventilatory endurance.[15] Thus, any treatment improving the MVV basically improves ventilatory muscle endurance. Consequently, the improvement observed after nutritional supplementation can be explained by an increase in respiratory muscle strength and endurance.

This is so perhaps because a balanced and enriched diet is required for an anabolic state of the body and for building the muscles in general and the ventilatory muscles in particular. Poorly nourished patients have high catabolism and they have low respiratory muscle strength as well as low level of endurance.

The ventilatory muscles can be trained for both strength and endurance in normal subjects[15], quadriplegic patients[16], patients with cystic fibrosis[17], and patients with COPD.[18,19] This treatment has also been shown to improve the exercise performance of patients with COPD.[18,19] The combined treatment of NS and RB was found to improve FVC similarly to the effect of nutritional supplementation only. However, it induced a greater change in MVV compared to the first two months of NS only. This can be explained by the fact that the ventilatory efficiency improved by both the increase in energy supplied to the ventilatory muscles and their work capacity, as shown by the following relationship.[26]

$$t_{lim} = \frac{\overline{\alpha E}}{W - \beta E}$$

t_{lim} = limited endurance time.
α = energy stored in the muscles.
E = muscle efficiency.
β = energy supplied to the muscles.
W = work done by the muscles

If βE is greater than the energy required for production of work, work could continue indefinitely. However if the βE is less than that required in order to produce work, there will be a limited endurance time. The combined treatment most likely improves the energy supplied to the ventilatory muscles on one hand and their work capacity and efficiency on the other.

Since only five patients were used in this study the results can be considered as preliminary and more work should be done to confirm these results as well as test the ventilatory muscle contractility and exercise capacity of these patients.

It was concluded that nutritional supplementation can improve ventilatory capacity and endurance, but a combined treatment of nutritional supplementation and resistive breathing training has a greater potential influence on the ventilatory capacity, specifically greater effect on the ventilatory endurance.

References

1. Rogers R., Dauner J., Sanders M., Claypool W.D., Openbrier D., Irwin M.: Nutrition and COPD: State of the art minireview. Chest. 1984; (suppl.) 85: 63-66
2. Braun S.R., Keim N.L., Dixon R.M., Lagnaz P., Anderegg A., Shargo E.S;: The prevalence and determinants of nutritional changes in chronic obstructive pulmonary disease. Chest. 1984; 87: 558-563
3. Hunter A.M.B., Carey M.A., Larsh H.W.: The nutritional status of patients with chronic obstructive pulmonary disease. Am. Rev. Resp. Dis. 1981; 124: 376-381
4. Vanderbergh E., Van De Voestign K.P., Gyslen: Weight changes in terminal states of chronic obstructive pulmonary disease. Am. Rev. Resp. Dis. 1967; 95: 556-566
5. Braun S.R., Dixon R.M., Keim N.L., et al.: Predictive clinical value of nutritional assessment factor in COPD. Chest. 1984; 85: 353-357
6. Driver A.G., McAlevy M.T.R., Smith T.L.: Nutritional assessment of patients with chronic obstructive pulmonary disease and acute respiratory failure. Chest. 1982; 82: 568-571
7. Hoch D., Murray D., Blalock et al.: Nutritional status as index of mortality in chronic airflow limitation. Chest 1984; 85: 66-67
8. Vandenbergh E., Van De Voestign K.P., Bilier et al.: Evaluation et propoistic de la bronchite chronique au stade de la retention de CO_2. Bull. Eur. Physiopathol. Respir. 1965; 1:260

9. Arora N.S., Rochester D.F.: Effect of body weight and muscularity on human diaphragm muscle mass, thickness and area. J. Appl. Physiol. 1982; 52:64-70

10. Arora N.S., Rochester D.F.: Respiratory muscle strength and maximum ventilation in undernourished patients. Am. Rev. Resp. Dis. 1982; 5-8

11. Sharp J.T.: The respiratory muscles in chronic obstructive pulmonary disease. Am. Rev. Resp. Dis. 1986; 136: 1089-1091

12. Roussos C., Fixley M., Gross D., Macklem P.T.: Fatigue of inspiratory muscles and their synergic behavior. J. Appl. Physiol. 1979; 5: 897-904

13. Rochester D.F., Braun N.M.T.: Determinants of maximal inspiratory pressures in chronic obstructive pulmonary disease. Am. Rev. Resp. Dis. 1985; 132:42-47

14. Wilson D.O., Rogers R.M., Sanders M.H., Pennock B.E., Reily J.J.: Nutritional intervention in malnourished patients with emphysema. Am. Rev. Resp. Dis. 1986; 134: 672-677

15. Leith D.E., Braddly M.: Ventilatory muscle strength and endurance training. J. Appl. Physiol. 1976; 41:508-516

16. Gross D., Ladd H.W., Riley E.J., Macklem P.T., Grassino A.: The effect of training on strength and endurance of the diaphragm in quadriplegia. Am. J. Med. 1980; 68: 27-35

17. Keens T.G., Krastins IRB, Wannamaker E.M., Levison H., Crozier D.N., Bryan A.C.: Ventilatory muscle endurance training in normal subjects and patients with cystic fibrosis. Am. Rev. Resp. Dis. 1977; 116: 853-860

18. Belman M., Mittman L.: Ventilatory muscle training improves exercise capacity in chronic obstructive pulmonary disease patients. Am. Rev. Resp. Dis. 1980; 121:273-280

19. Pardy R.L., Rivington R.N., Despas P.J., Macklem P.T.: Muscle training compared with physiotherapy in patients with chronic airflow limitation. Am. Rev. Resp. Dis. 1981; 123: 421-425

20. Monod H., Scherrer J.: The work capacity of synergic muscular group. Ergonomics 1965; 8: 329-337

21. Jardim J., Farkas G., Prefault C., Thomas D., Macklem P.T., Roussos C.S.: The failing inspiratory muscles under normoxic and hypoxic condition. Am. Rev. Resp. Dis. 1981; 124: 264-66

22. Keys A., Fidanza F., Karvonen M.J., Kimura N., Taylor H.I.: Indices of relative weight and obesity. J. Chronic. Dis. 1972; 25: 329-343

23. Shils M.E., Young V.R.: *Modern nutrition in health and disease, etiology.* Philadelphia, Lea and Febiger, 1988; 831-885

24. Morris J.F., Kosk A., Johnson L.C.: Spirometric standards for healthy nonsmoking adults. Am. Rev. Resp. Dis. 1971; 103:57-67

25. Wouters E.F.M., Schols A.M.W.: Cachexia in chronic obstructive pulmonary disease (COPD). This volume p. 167

Subject Index

Fibronectin   13
Functional residual capacity (FRC)   67, 202

G
Gallbladder disease   65
Gamma linolenic acid (DHLA)   178
Gastric colonization   95
GI bleeding, osteomalacia   69
Glucose system   154
Glycogen   28
Gout   65
Gram-negative organisms   95

H
Hand grip dynamometry   13
Healthy subjects   24
Hepatic steatosis   58
Hiatus hernia   65
Home parenteral nutrition (HPN)   176
Hormones   143
Hospitalization   18, 80
Hospitalized COPD patients   29, 37, 42
Human kwashiorkor   92
Humoral immune function   94
Hyperalimentation   138
Hypercapnia   66, 186
Hypercatabolism   168
Hyperglycemia   144
Hyperinflation   197
Hypermetabolic patients   47, 159, 160
Hypermetabolism   31, 48, 80, 168
Hyperpnoea   100, 101
Hypersomnolence   66
Hypertension   58, 65
Hyperthyroidism   48
Hypertriglyceridemia   58
Hyperventilation   49
Hypogonadism   65
Hypokalaemia   82
Hypomagnesiemic patients   77, 81
Hypophosphataemia   81, 144
Hypoxaemia   68, 184
Hypoxia   54, 55, 80